AF564548

Fundamentals of Biochemistry

NIPA® GENX ELECTRONIC RESOURCES & SOLUTIONS P. LTD.
New Delhi-110 034

About the Editors

Dr. Aarif Ali Division of Veterinary Biochemistry, Faculty of Veterinary Sciences and Animal Husbandry SKUAST-K, Shuhama, Alusteng, Srinagar-190006, J&K, India Dr. Aarif Ali holds Doctor of Philosophy in Clinical Biochemistry at the University of Kashmir and has collaboratively worked and published more than 80 research publications in different reputed National as well as International journals. He has obtained his BSc from SP College, affiliated with University of Kashmir, and Masters in Clinical Biochemistry from the University of Kashmir (Jammu and Kashmir, India), respectively. Dr. Ali has worked as Senior Research fellow in national (DST & ICMR) sponsored research projects. Dr. Aarif Ali serves as reviewer of several high-impact national as well as international scientific journals. Dr. Ali has recently published books on Pharmacogenomics From Discovery to Clinical Implementation (Elsevier, ISBN 978-0-443-15336-5) and Biological Insights of Multi-Omics Technologies In Human Diseases (Elsevier, ISBN 978-0-443-23971-7).

Dr. Ali is engaged in studying various horizons of biochemistry, molecular biology, clinical diagnosis, bioinformatics and its application in human health with a keen interest in free radical biology, oxidative stress, inflammation, diabetes, cancer biology, therapeutic targets, and medicinal plant based research.

Dr. Ali's contributions to the scientific community extend to numerous research publications, covering various aspects of human health. Additionally, He has pursued continuous professional development through a range of training and certifications, showcasing his commitment to staying at the forefront of his field. Dr. Ali's dedication to research and his passion for contributing to scientific knowledge make him a valuable asset to the academic and scientific community.

Dr. Aadil Ayaz Hemvati Nandan Bahuguna Garhwal University, Srinagar, Uttarakhand, India Dr. Aadil Ayaz is an accomplished Biotechnologist and Molecular Biologist with good academic background and extensive research experience. Dr. Aadil earned doctorate in Biotechnology with a special focus on animal fibre omics, from Hemvati Nandan Bahuguna Garhwal University, Srinagar, Uttarakhand. Dr. Aadil worked in collaboration with my reputed state and national institutions. He has published over 50 research articles in different reputed national and international scientific journals and serves as a reviewer for many as well. Dr. Aadil has worked as Senior Research fellow in national (DBT, ICAR) and state (JKCST) sponsored research projects. Currently Dr. Aadil is engaged in studying the Diagnosis, Transmission and management of Virus in temperate fruits. Prior to this, Dr. Aadil served as a research scientist in the Viral Research and Diagnostic Laboratory at SKIMS Medical College Srinagar. Dr. Aadil is also involved in studying various horizons of biotechnology, biochemistry, bioinformatics and their application in human welfare/ health, with a special focus on cancer biology, therapeutics targets and medicinal plant based research.

Dr. Sheikh Bilal Ahmad Professor Division of Veterinary Biochemistry, Faculty of Veterinary Sciences & Animal HusbandrySKUAST-K, Shuhama, Alusteng, Srinagar, J&K, India-190006 Dr. Sheikh Bilal Ahmad is a Senior faculty member at the Division of Veterinary Biochemistry, Faculty of Veterinary Sciences & Animal Husbandry SKUAST-K, Shuhama, Alusteng, Srinagar, J&K, India. Dr. Sheikh Bilal Ahmad holds Doctorate in Biochemistry from Department of Biochemistry and Bioprocess Technology, Sam Higginbottom Institute of Agriculture, Technology and Sciences Allahabad, India. Dr. Sheikh Bilal Ahmad has more than twenty years of research and teaching experience in veterinary biochemistry, cancer biology, molecular biology and genetics. He has published more than 100 research papers in peer-reviewed, international journals, and has more than 30 book chapters. Dr. Sheikh Bilal Ahmad serves as an editorial board member and reviewer of several high-impact, international scientific journals.

Fundamentals of Biochemistry

Aarif Ali
Division of Veterinary Biochemistry
Faculty of Veterinary Sciences & Animal Husbandry
Sher-e-Kashmir University of Agricultural Sciences and Technology-K
Shuhama, Alusteng, Srinagar, J&K, India-190006

Aadil Ayaz
Hemvati Nandan Bahuguna Garhwal University
Srinagar, Uttarakhand, India

Sheikh Bilal Ahmad
Division of Veterinary Biochemistry
Faculty of Veterinary Sciences & Animal Husbandry
Sher-e-Kashmir University of Agricultural Sciences and Technology-K
Shuhama, Alusteng, Srinagar, J&K, India-190006

NIPA® GENX ELECTRONIC RESOURCES & SOLUTIONS P. LTD.
New Delhi-110 034

**NIPA® GENX ELECTRONIC
RESOURCES & SOLUTIONS P. LTD.**

101,103, Vikas Surya Plaza, CU Block
L.S.C. Market, Pitam Pura, New Delhi-110 034
Ph : +91-11-43860225, Mob.: +91 9717133558, 9540816132
E-mail: newindiapublishingagency@gmail.com
Website: www.nipaersources.com

Print ISBN: 978-93-58872-68-2
ebook ISBN: 978-93-58878-13-4

Composed and Designed by NIPA®.

Preface

Biochemistry serves as the vital link between biology and chemistry, unraveling the intricate molecular mechanisms that underlie life itself. This comprehensive book covers a wide range of topics within the ever-evolving field of biochemistry. Our aim is to provide a thorough introduction to these fundamental concepts, while also highlighting the latest advancements and their implications for the fields of science and medicine. The motivation behind this book is to create a resource that is accessible to newcomers, yet still valuable to those with a deeper understanding of the subject. Our ultimate goal is to present fundamental concepts of basic biochemistry in a clear, engaging, and systematic manner. To achieve this, we have incorporated various pedagogical tools, including detailed illustrations, step-by-step explanations, and real-world examples that demonstrate the relevance of biochemistry in everyday life. The book is organized into several sections, each focusing on a key area of biochemistry. We begin by exploring the basic building blocks of life, such as amino acids, nucleotides, carbohydrates, and lipids, before delving into their assembly into larger macromolecules like proteins and nucleic acids. Subsequent chapters delve into the dynamic aspects of basic biochemistry, enzyme kinetics and vitamins. The invaluable feedback from our colleagues and students, as well as their unwavering enthusiasm for biochemistry, has been a constant source of inspiration. We would also like to acknowledge the unwavering support of our families, whose patience and encouragement have been instrumental in the completion of this project. We envision this book as a valuable resource for students, educators, and researchers. Whether you are new to the field of biochemistry or looking to expand your knowledge, we are certain that you will find this text to be both educational and inspiring. This template is flexible and can be tailored to suit the specific content and emphasis of your biochemistry book.

Editors

Contents

1

General Biochemistry

Introduction and Scope of Biochemistry

Biochemistry is the field of study that focuses on studying chemical reactions occurring within living organisms. It is a specialized branch that combines the principles of both chemistry and biology. In order to comprehend the molecular mechanisms behind every biological event, this multidisciplinary field incorporates elements of biology, chemistry, and physiology. The scope of biochemistry extends from the study of small molecules and their interactions within cells to the analysis of complex metabolic pathways, enzyme kinetics, and the structure and function of biomolecules.

One of the main goals of biochemistry is to elucidate the chemical reactions that take place in living systems and understand how they regulate key biological processes. Biochemists investigate the structure, function, and properties of various biomolecules, such as proteins, carbohydrates, lipids, and nucleic acids. It helps us to study how these biomolecules interact and contribute to the overall functioning of cells, tissues, and organisms.

Biochemistry also plays a vital role in various areas of research and practical applications. For instance, in medical biochemistry, scientists study the molecular basis of diseases to develop diagnostic tests and therapeutic interventions. They investigate the biochemical mechanisms underlying genetic disorders, cancer, metabolic, and infectious diseases. Thus, understanding the biochemical basis of diseases can lead to the development of targeted and personalized treatments.

Industrial applications of biochemistry are numerous, and they also play a significant role in genetic engineering and biotechnology. It is used in the production of pharmaceuticals, biofuels, and agricultural products. By manipulating the genetic material of organisms, new enzymes, optimized metabolic pathways, and recombinant proteins on a large scale can be produced.

The scope of biochemistry is also closely linked to other scientific disciplines. For example, it intersects with molecular biology, which focuses on studying genes and their regulation. It overlaps with cell biology, exploring how cells function at the molecular level. Additionally, biochemistry has strong connections with medicine, pharmacology, and toxicology, as understanding the molecular basis of diseases and drug actions is crucial in these fields.

In conclusion, biochemistry encompasses a wide range of scientific inquiries about the chemical processes happening within living organisms. It investigates the structure, function, and interactions of biomolecules, and its scope extends from fundamental research in molecular biology to practical applications in medicine, industry, and biotechnology. By uncovering the molecular mechanisms behind life processes, biochemists make significant contributions to human health, agriculture, and the development of new technologies.

Applications of Biochemistry

Biochemistry has a wide range of applications in various fields. In this article, we will delve into the applications of biochemistry and how it impacts our lives.

One of the significant applications of biochemistry is in the field of medicine. Biochemical research has helped to elucidate the molecular basis of diseases, leading to the development of diagnostic tools and therapeutic interventions. For instance, the discovery of specific biochemically altered molecules in the blood or urine has revolutionized the diagnosis of various conditions, such as diabetes or certain types of cancer. Biochemical techniques, such as enzyme-linked immunosorbent assay (ELISA), polymerase chain reaction (PCR), and mass spectrometry, have become essential tools in medical laboratories worldwide. Moreover, biochemistry has paved the way for the production of life-saving drugs. Pharmaceutical companies rely on studying the biochemical pathways within the body to design and develop effective medications. By understanding the biochemical targets and metabolic processes, scientists can create drugs that specifically interact with enzymes, receptors, or other molecules involved in disease progression.

Another significant area where biochemistry finds applications is in agriculture and food science. Biochemical research helps improve crop yield, develop disease-resistant plants, and increase the nutritional value of food. Scientists can optimize fertilizer compositions and develop strategies to combat crop diseases by studying plant metabolism. Additionally, biochemistry is crucial in food preservation and processing techniques, ensuring food safety and quality. Additionally, biochemistry plays a vital role in environmental sciences and biotechnology. Understanding the biochemical processes within ecosystems allows scientists to evaluate the impact of pollutants and develop strategies for environmental remediation. Biochemical techniques are also widely used in biotechnology, enabling the manipulation of genes and proteins for various applications, such as producing biofuels, developing genetically modified organisms (GMOs), and manufacturing pharmaceuticals through bioprocessing.

Biochemistry also extends its applications to forensic science. Forensic scientists can identify individuals, determine paternity, and provide valuable evidence in criminal investigations by analyzing DNA, proteins, and other biochemical markers. The field of forensic toxicology utilizes biochemistry to detect and analyze drugs, poisons, and other substances in biological samples, helping to establish the cause of death in cases of suspected poisoning or drug overdose.

In general, biochemistry has a vast scope of applications in various fields. Biochemistry is crucial in advancing our understanding of living organisms and their complex chemical processes from medicine to agriculture, environmental sciences to forensic science. The findings and techniques developed in biochemistry have profoundly impacted human health, agriculture, the environment, and forensic investigations.

As technology and our understanding of biochemistry continue to advance, we can expect even more exciting applications and discoveries in the future.

Transport across the cell membrane

The cell membrane, also called the plasma membrane, is a phospholipid bilayer that plays a crucial role in controlling the movement of molecules. Its primary function is establishing the cell's border and maintaining its functionality. The plasma membrane exhibits selective permeability, allowing certain substances to enter or exit the cell freely. In contrast, others necessitate the utilization of specialized transport proteins and occasionally even energy expenditure to facilitate their passage. The transport of molecules across the cell membrane depends upon the molecule's size and hydrophobicity. Transport proteins or permeases are of two types, i.e., carriers and ion channels. In active and passive transport, carrier molecules are involved, whereas ion channels participate in the passive mode of transport. Furthermore, carrier transport proteins can be further classified based on the orientation of the molecules being transported. In cotransport, two substances are transported together in the same direction across a cell membrane. As a result, the following two classes, i.e., uniport and co-transport (antiport and symport), exist.

a) **Uniport:** Uniport refers to the transportation of a solitary molecule in a single direction, independent of any concurrent transportation of other molecules or ions. This mode of transportation necessitates a uniporter, an integral membrane protein, such as an ion channel or carrier protein. Glucose transporters, known as GLUTs, function as uniporters responsible for the uptake of glucose.

b) **Antiport:** In antiport, two molecules are transported simultaneously in opposite directions, carried by an antiporter transmembrane protein molecule. The sodium-potassium pump (Na^+/K^+ pump) is a prime illustration of this type.

c) **Symport:** Symport refers to the transportation of two molecules simultaneously in the same direction. A prime illustration of this type is the sodium-glucose pump (Na^+-glucose pump). In mammals, glucose is transported through sodium-dependent glucose transporters, which require energy for this process. As glucose and sodium are both transported in the same direction across the membrane, these transporters are categorized as symporters.

Based on the nature of the transport through the plasma membrane, there are two categories

1. Passive transport
2. Active transport

Passive transport

The transport of substances through the cell membrane via passive transport along a gradient is an energy-free mechanism. It occurs spontaneously without the cell needing to utilize any of its energy resources. During passive transport, substances migrate from regions of higher concentration to lower concentration, following the concentration gradient present in the given space. Polar or massive substances diffuse independently across the lipid bilayer via membrane proteins such as transporters, permeases, carriers, and channels. The four major types of passive transport are simple diffusion, facilitated diffusion, osmosis, and filtration.

Types of passive transport

1. Simple Diffusion

Diffusion is a phenomenon characterized by the movement of a substance from an area of greater concentration to an area of lesser concentration without the involvement of any external transport protein, thus leading to the equalization of concentration levels. This spontaneous process occurs without the need for any external energy input and is therefore referred to as passive transport. Non-polar molecules, including O_2, CO_2, and N_2, can readily undergo diffusion across a membrane, traversing from regions of higher concentration to regions of lower concentration.

2. Facilitated diffusion

Facilitated diffusion is a form of passive transport where molecules traverse the cell membrane rapidly, facilitated by specific permeases present in the membrane. This process exclusively takes place in the direction of a concentration gradient and does not necessitate metabolic energy. To meet the cellular requirements, ions such as chloride ion sodium (Na^+), calcium (Ca^{2+}), potassium (K^+), (Cl^-), bicarbonate ion ($HCO3^-$), and polar molecules like glucose molecules are transported via facilitated diffusion. The movement of water-mediated by protein channels called aquaporin's is also an illustration of facilitated diffusion.

3. Filtration

Filtration represents another passive process of transporting materials across a cell membrane. While diffusion and osmosis depend on concentration gradients, filtration operates through a pressure gradient. Molecules move from regions of higher pressure to regions of lower pressure. Filtration is non-selective, meaning it does not sort molecules; they pass through the membrane due to pressure gradients and their size. If molecules are sufficiently small, they can traverse the membrane. The force responsible for propelling these molecules is referred to as hydrostatic pressure. The capillary process exchange relies heavily on filtration as one of its primary methods. Through the hydrostatic pressure force generated by blood pressure, substances are compelled to exit capillaries and enter the tissue fluid that envelops cells. Moreover, this mechanism is also responsible for generating the filtrate that initiates urine formation within the kidney.

4. Osmosis

Osmosis is the process by which water diffuses through a semipermeable membrane from higher to lower water potential gradient. This diffusion is influenced by the solute concentration gradient, which refers to the difference in water levels on either side of the membrane. The concentration of solutes is inversely related to the amount of water present in a solution. In simpler terms, when the water level is high, the solute concentration is low, and vice-versa, it forms three different solutions, i.e. isotonic, hypertonic, and hypotonic.

Isotonic solution

An isotonic solution refers to a solution that contains an equal concentration of solutes inside and outside the cell. The normal functioning of blood cells relies on the presence of an isotonic solution in the plasma surrounding them, as opposed to the solution inside the cells. This isotonic solution facilitates the movement of water and nutrients into and out of the cells. It is vital for blood cells to fulfill their duty of delivering oxygen and other nutrients to different body areas.

Hypertonic solution

In the case of a hypertonic solution, the solute concentration outside the cell surpasses that inside the cell. Consequently, blood cells exposed to a hypertonic environment undergo plasmolysis, resulting in insufficient water to support cellular functions.

Hypotonic solution

When the solute concentration inside a cell exceeds that of the outside the cell, it is referred to as a hypotonic solution. In the presence of a hypotonic environment, cells will undergo lysis, causing their contents to be released into the bloodstream. This can have detrimental consequences, including the loss of numerous blood cells and the occurrence of dangerous side effects.

Examples of osmosis

The significance of osmosis extends to plants, animals, and humans. A few examples of osmosis are as below.

1. The transfer of water from the soil into the plant roots is facilitated by osmosis. This occurs because the solute concentration in the roots is higher than that of the surrounding soil, causing water to flow into the roots.
2. Osmosis also impacts individuals afflicted with cholera. The proliferation of bacteria in the intestines disrupts the natural flow of water absorption, causing dehydration.
3. When a fish, whether freshwater or saltwater, is exposed to water with varying salt concentrations, its cells are affected by the influx or efflux of water, ultimately resulting in the fish's demise.

Types of osmosis

There are two types of osmosis, i.e., endosmosis and exosmosis.

Endosmosis

Endosmosis describes the phenomenon that occurs when a substance or a cell is placed in a hypotonic solution, causing the solvent molecules to penetrate the cell. This influx of solvent leads to the cell becoming turgid or undergoing deplasmolysis. Hence, it is referred to as endosmosis.

Exosmosis

Exosmosis refers to the process where a substance or a cell, when exposed to a hypertonic solution, causes the solvent molecules to exit the cell, resulting in the cell becoming flaccid or undergoing plasmolysis.

Active transport

Active transport involves the transportation of molecules or ions across a cell membrane, moving from an area of lower concentration to an area of higher concentration. This process requires cellular energy and goes against the concentration gradient. Membrane proteins facilitate active transport. Active transport can be categorized as primary or secondary based on the energy source.

Primary active transport

It is also known as direct transport, as energy is derived from ATP and light, which drive the transport. The nature of the primary transport may be uniporter or co-transporter. The Sodium-potassium pump is an example of primary active transport.

Secondary active transport

It is also known as indirect transport, in which an existing electrochemical gradient provides the necessary energy. The nature of secondary active transport is co-transport. Na^+-glucose symporter serves as an illustration of secondary transport.

Bulk transport

Bulk transport serves to transport substantial quantities of materials or food particles across the cell membrane. This process occurs in both inward and outward directions and relies on the assistance of carrier molecules. This indispensable biological process facilitates the efficient movement of significant amounts of substances, including nutrients, water, and waste products. There are two distinct types of bulk transport, i.e., endocytosis and exocytosis, which both necessitate energy utilization in the form of ATP.

Endocytosis

The process by which cells engulf and absorb larger molecules or particles from their surroundings is known as endocytosis. During this process, the plasma membrane envelops the substance to be internalized and buds off inside the cell. It is of two types: phagocytosis and pinocytosis.

Phagocytosis

It is also known as cell eating, and during this process, cells take up and consume solid particles. In the human defense system, it is an essential process as various cells known as phagocytes kill or engulf the foreign material or particle. Protists are among the unicellular creatures that use this specific mechanism for eating. In amoeba, the process of phagocytosis involves the engulfment of nutrients by pseudopods located throughout the cell; in ciliates, on the other hand, the process occurs within a specialized chamber or groove called the cytostome.

Pinocytosis

It is also known as cell drinking and is a form of endocytosis in which the cells engulf and internalize extracellular fluid and dissolved molecules. In pinocytosis, tiny droplets of fluid are taken up by the cell through invaginations of the cell membrane, forming small vesicles. This process enables the cell to obtain nutrients, molecules, and ions from its surroundings. Pinocytosis plays a crucial role in various physiological functions, including nutrient absorption in the intestine and immune system functions.

Exocytosis

Exocytosis is the fundamental process that involves the release of substances or molecules from the cell into the extracellular space, playing a crucial role in numerous physiological processes, including neurotransmission, hormone secretion, immune response, and cell-to-cell communication.

Types of exocytosis

The process of exocytosis is generally categorized into two groups, namely constitutive and regulated exocytosis

1. Constitutive exocytosis

Constitutive exocytosis is a ubiquitous mechanism occurring in all cells, which involves the uninterrupted release of newly synthesized protein cargo. This vital pathway facilitates the transportation of proteins and lipids to the cell membrane. Constitutive exocytosis does not necessitate any specific signal for the secretion of materials outside the cell. Notably, eosinophils and other cell types possess the inherent ability to produce and secrete extracellular matrix proteins continuously.

2. Regulated exocytosis

Regulated exocytosis is a vital process that allows secretory cells to transport molecules to the cell surface and the extracellular space. In response to stimulation, this process involves the fusion of cytoplasmic organelle membranes with the plasma membrane, facilitating the release of molecules. Regulated exocytosis is a highly regulated process that is activated by calcium and serves as the foundation for numerous intercellular signaling mechanisms. Several instances of regulated exocytosis include the secretion of hormones, enzymes, and neurotransmitters. Pancreatic beta cells are specialized cells that produce insulin, a hormone secreted into the bloodstream when blood sugar levels rise. Moreover, within the neuronal cells, specific vesicles are formed containing neurotransmitters as secreted via regulated exocytosis during the process of synapsis or neurotransmission.

Buffers

A buffer solution is characterized by its ability to withstand significant changes in pH when small amounts of acid or base are added to it. Typically, buffer solutions comprise a weak acid and its corresponding conjugate base or a weak base and its corresponding conjugate acid, enabling them to maintain a stable pH. Buffering is essential in numerous chemical reactions that necessitate a constant pH. In nature, various systems employ buffering to regulate pH levels. For instance, the bicarbonate buffering system plays a crucial role in regulating blood pH, and bicarbonate also acts as a buffer in the ocean.

Characteristics of buffer solution

1. Buffer solutions possess distinct pH values that remain constant over time, even when left undisturbed for extended periods.
2. The pH of buffer solutions remains unchanged upon dilution and prolonged standing, showcasing their stability.
3. Buffer solutions exhibit remarkable resistance to changes in pH, even adding minute amounts of acids or bases at a constant temperature.

Types of buffer solutions

These solutions are vital in maintaining a constant pH level, ensuring the stability and functionality of various chemical reactions and biological systems. The classification of buffer solutions primarily revolves around two types: acidic buffers and alkaline buffers.

Acidic buffers

As their name suggests, acidic buffers are crucial in maintaining acidic conditions. These buffers are prepared by combining a weak acid with its corresponding salt with a strong base. The pH of these solutions is below 7. A prime example of an acidic buffer solution is the mixture of sodium acetate and acetic acid, resulting in a pH value of 4.74.

A mixture of CH_3COOH (weak acid) and CH_3COONa (salt) is present in an aqueous solution.

$CH_3COOH \leftrightharpoons CH_3COO^- + H^+$ (incomplete dissociation)

$CH_3COONa \rightarrow CH_3COO^- + Na^+$ (complete dissociation)

$H_2O \leftrightharpoons H^+ + OH^-$ (incomplete dissociation)

When a drop of strong acid (HCl) is added to the aforementioned buffer solution, the H^+ ions react with the CH_3COO^- ions, forming weakly ionized CH_3COOH. However, the ionization of CH_3COOH is further hindered due to the common ion effect. Consequently, the pH of the solution remains unaffected.

Upon introducing a strong base (NaOH) drop to the buffer solution described above, it (NaOH) reacts with the free acid, forming undissociated water molecules. Consequently, the pH of the solution remains unaffected.

$CH_3COOH + NaOH \leftrightharpoons CH_3COONa + H_2O$

Basic buffer

These particular solutions serve the purpose of maintaining basic conditions. A basic buffer solution is created by combining a weak base and its corresponding salt with a strong acid, resulting in a solution with a basic pH. The pH value of these solutions exceeds 7. In an aqueous solution containing equimolar quantities of ammonium hydroxide and ammonium chloride, the pH level measures 9.25.

In an aqueous solution, a mixture of NH_4OH (weak base) and NH_4Cl (salt) are combined in basic buffers.

$NH_4OH \leftrightharpoons NH_4^+ + OH^-$ (incomplete dissociation)

$NH_4OH + HCl \leftrightharpoons NH_4Cl + H_2O$

$NH_4Cl \rightarrow NH_4^+ + Cl^-$ (complete dissociation)

$H_2O \leftrightharpoons H^+ + OH^-$ (incomplete dissociation)

Upon the addition of HCl, the H^+ ions react with NH_4OH, leading to the formation of undissociated water molecules. Consequently, the pH of the buffer remains unaltered.

$NH_4OH + OH^- \leftrightharpoons NH_4^+ + H_2O$

Upon the addition of NaOH, the OH^- ions react with $NH4^+$ ions, leading to the formation of NH_4OH, which exhibits weak ionization. The presence of a common ion further diminishes this reaction. Consequently, the pH of the buffer remains unaltered.

Applications of buffers

1. Buffers play a crucial role in a wide range of industrial processes, including the production of paper, dyes, inks, paints, and pharmaceuticals. They provide stability and control over pH levels, ensuring optimal conditions for these manufacturing processes.
2. Besides their industrial applications, buffers are extensively utilized in agriculture, dairy products, and the preservation of various foods and fruits. By maintaining the pH balance, buffers help enhance the quality and shelf life of these products, ensuring they remain fresh and safe for consumption.
3. Indicators rely on buffers to accurately determine the pH of a solution. Whether in a laboratory setting or in everyday applications, buffers provide the necessary stability to ensure precise pH measurements, enabling scientists and researchers to gather reliable data for their experiments and analyses.
4. The human body relies on blood as a natural buffer system. Maintaining a pH of 7.4 in blood plasma is vital for sustaining life, as enzyme catalysis is highly sensitive to pH levels. The body's ability to regulate pH ensures optimal functioning of enzymes and overall physiological processes.
5. In qualitative inorganic analysis, buffers such as CH_3COOH+ CH_3COONa are employed after the second group to remove phosphate ions. This buffer system aids in accurately identifying and separating different ions, facilitating the analysis and characterization of various substances.

Henderson-Hasselbalch Equation

The Henderson-Hasselbalch equation serves as a fundamental tool in understanding the interplay between the pH of acids in aqueous solutions and their acid dissociation constant (pKa). This equation enables the estimation of the pH of a buffer solution, provided that the concentrations of either the acid and its conjugate base or the base and its corresponding conjugate acid are known. Lawrence Joseph Henderson, an American chemist, is credited with the initial derivation of the equation. Later on, Karl Albert Hasselbalch, a chemist from Denmark, provided an alternative formulation of the equation utilizing logarithmic terms. The Henderson-Hasselbach equation is written as:

$$pH = pK, + \log \frac{[\text{conjugate base}]}{\text{acid}}$$

OR

$pH = pK_a + \log_{10}([A^-]/HA])$ where

$[A^-]$ = molar concentration of conjugate base of the acid

[HA] = molar concentration of weak acid

Derivation of equation

The equation for the ionization of a weak acid HA can be derived by considering an example as mentioned below:

$$HA + H_2O \leftrightharpoons H^+ + A^-$$

Its acid dissocation constant (Ka) can be written as follow:

$Ka = [H^+] [A^-] [HA]$

Now taking the negative log of both sides:

$-\log Ka = -\log [H^+] [A^-] [HA]$

$$\Rightarrow -\log Ka = -\log[H^+] - \log [A^-][HA]$$

we know that $-\log [H^+] = pH$

and

$-\log Ka = pKa$

Moreover, the above equation can also be written as,

$pKa = pH - \log [A^-] [HA]$

The equation on rearranging

$$\Rightarrow pH = pKa + \log[A^-][HA]$$

The equation known as the Henderson-Hasselbalch equation or Henderson equation is the designated nomenclature for this mathematical expression.

Assumptions

The Henderson-Hasselbalch equation is derived based on several assumptions. These assumptions are as follows:

Assumption 1

The acid HA is considered to be monobasic. This means it donates only one proton (H^+) when it dissociates in water.

Assumption 2

The self-dissociation of water is neglected. This means that the concentration of H^+ and OH^- ions resulting from the dissociation of water is assumed to be negligible compared to the concentration of the acid and its conjugate base.

Assumption 3

The salt formed from the reaction between the acid and a base is assumed to be completely dissociated in the solution. This means that all the ions present in the salt are considered to be fully separated and independent in the solution.

These assumptions simplify the mathematical derivation of the Henderson Hasselbalch equation and allow for a more practical application in various fields of chemistry and biology.

Limitations of Henderson-Hasselbalch Equation

Due to its assumption that the concentration of the acid and its conjugate base will remain constant at chemical equilibrium, the Henderson-Hasselbalch equation falls short in accurately predicting values for strong acids and strong bases. Furthermore, the equation's disregard for the binding of protons to the base and the self-dissociation of water leads to inaccurate pH values for highly diluted buffer solutions.

Buffer systems

A buffer system can withstand alterations in pH even when acid or base is added. The buffer systems within the human body exhibit remarkable efficiency, with each system operating at its own unique pace. The chemical buffers in the blood swiftly adapt to changes in pH, accomplishing this task within seconds. Three main buffering systems in the body are crucial in maintaining pH balance. These include the Bicarbonate buffer system, the phosphate buffer system, and the protein buffer system.

Bicarbonate buffer system

The bicarbonate buffer system is a highly significant and efficient buffering system that maintains the acid-base balance of the extracellular fluid (ECF) in the human body, particularly in the bloodstream, and it possesses a remarkable 53% buffering capacity. Consequently, the bicarbonate buffer system in the body is referred to as an open buffering system. It comprises carbonic acid, a weak acid, and the bicarbonate anion, its conjugate base. It involves the regulation of carbonic acid (H_2CO_3), bicarbonate ion ($HCO3^-$), and carbon dioxide (CO_2) levels to ensure proper pH in the blood and duodenum, as well as other tissues to support normal metabolic function. This process is facilitated by an enzyme called carbonic anhydrase, which catalyzes the reaction between carbon dioxide (CO_2) and water (H_2O) to form carbonic acid (H_2CO_3). The carbonic acid then rapidly dissociates into a bicarbonate ion ($HCO3^-$) and a hydrogen ion (H^+) by carbonic anhydrase. Like any other buffer system, the bicarbonate buffer system maintains pH balance by having a weak acid (H_2CO_3) and its conjugate base ($HCO3^-$) present, neutralizing any excess acid or base introduced into the system.

$CO_2 + H_2O \rightleftarrows H_2CO_3 \rightleftarrows HCO3^-$

If this system fails to function properly, it can lead to an imbalance in the acid-base levels, resulting in conditions like acidemia (pH < 7.35) or alkalemia (pH > 7.45) in the blood.

Phosphate buffer system

Despite its limited significance as an extracellular fluid buffer, the phosphate buffer system is vital in buffering renal tubular fluid and intracellular fluids (ICF). Composed of two primary constituents, namely the weak acid monosodium dihydrogen phosphate (NaH_2PO4^-) and its conjugate base disodium hydrogen phosphate ($Na_2HPO_4{}^{2-}$), phosphate buffers exhibit distinct behavior when exposed to strong acids or bases. In the presence of a strong acid such as HCl, $Na_2HPO_4{}^{2-}$ acquires an additional hydrogen ion, forming the weak acid $Na_2H_2PO_4{}^-$ and sodium chloride (NaCl). Conversely, when $Na_2HPO_4{}^{2-}$ (the weak acid) encounters a strong base like sodium hydroxide (NaOH), it reverts back to the weak base and generates water. Although acids and bases persist, they retain their respective ions.

$HCl + Na_2HPO4 \rightarrow NaH2PO_4 + NaCl$

(strong acid) + (weak base) → (weak acid) + (salt)

$NaOH + NaH_2PO_4 \rightarrow Na_2HPO_4 + H_2O$

(strong base) + (weak acid) → (weak base) + (water)

The phosphate buffer system, with a pK of 6.8, operates near its maximum buffering power due to its proximity to the normal pH of 7.4 in body fluids. However, its concentration in the extracellular fluid is significantly lower, only about 8 percent of the bicarbonate buffer concentration. As a result, the total buffering power of the phosphate system in the extracellular fluid is much less than that of the bicarbonate buffering system. In contrast, the phosphate buffer plays a crucial role in the tubular fluids of the kidneys for two reasons. Firstly, phosphate becomes highly concentrated in the tubules, thereby increasing the buffering power of the phosphate system. Secondly, the tubular fluid generally has a lower pH than the extracellular fluid, which brings the buffer's operating range closer to its pK of 6.8. Moreover, the phosphate buffer system is essential in buffering intracellular fluid as phosphate concentration in this fluid is significantly higher than in the extracellular fluid. Additionally, the pH of intracellular fluid is lower than that of extracellular fluid, making it closer to the pK of the phosphate buffer system compared to the extracellular fluid.

Protein buffer system

The protein buffer system plays a crucial role in maintaining the acidity levels both inside and around cells. Proteins, composed of amino acids, possess amino groups with positive charges and carboxyl groups with negative charges. These charged regions enable the proteins to bind hydrogen (H^+) and hydroxyl (OH^-) ions, thereby functioning as buffers. In the blood, protein buffers regulate the pH by absorbing and releasing excess hydrogen ions, ensuring a balanced pH level. The blood contains protein buffers, including hemoglobin (150 g/l) and plasma proteins (70 g/l).

The hemoglobin buffer system in red blood cells is a vital protein responsible for regulating blood pH levels by specifically buffering CO_2. Its mechanism involves the binding and release of protons (H^+) in response to alterations in blood pH. In the event of an increase in blood pH, hemoglobin releases H^+ ions from multiple atomic sites within its structure. Conversely, hemoglobin binds to H^+ ions when blood pH decreases to counteract the pH reduction. Hemoglobin contains approximately three times the amount of histidine residues per molecule. The imidazole group of the histidine residues, with a pKa of around 6.8, plays a crucial role in the buffering process. This pKa value is well-suited for effective buffering at the physiological pH. Regarding quantitative importance, hemoglobin surpasses plasma proteins by approximately six times. This is due to its higher concentration and the presence of three times the number of histidine residues per molecule. For example, if the blood pH shifted from 7.5 to 6.5, hemoglobin would buffer 27.5 mmol/l of H^+, while the total plasma protein buffering would only account for 4.2 mmol/l of H^+. Hemoglobin, particularly in its deoxyhemoglobin form, functions as the most significant intracellular buffer, exhibiting enhanced effectiveness. Deoxyhemoglobin refers to hemoglobin that has released the oxygen it previously carried. Deoxyhaemoglobin proves to be a more efficient buffer than oxyhemoglobin, contributing to about 30% of the Haldane effect. The primary factor responsible for the Haldane effect in CO_2 transport is the significantly greater ability of deoxyhemoglobin to form carbamino compounds.

Summary

- Biochemistry involves the utilization of chemical principles to investigate biological phenomena occurring within cells and molecules.
- Active and passive transport are the two primary transportation methods across the membrane.
- Within cellular transport, active transport is crucial in facilitating the movement of substances like amino acids, glucose, and ions across cell membranes. This process specifically targets regions that already possess a high concentration of these substances.

- The entry of water molecules into the cell is called endosmosis, while the exit of water molecules from the cell is termed exosmosis.
- Active transport proteins, referred to as pumps, function by working against electrochemical gradients.
- ATP is crucial for primary active transport, as it relies directly on this energy source. On the other hand, secondary active transport operates differently as it utilizes electrochemical gradients and does not directly depend on ATP.
- Endocytosis refers to the cellular mechanism through which cells engulf and absorb larger molecules and particles from their surrounding environment.
- A buffer is a solution that can withstand changes in pH when acidic or basic substances are introduced.
- When a solution has a pH value below 7, it is referred to as acidic, while a solution with a pH value above 7 is considered basic.
- Buffer solutions, composed of a weak acid and its conjugate base, are particularly known for their ability to resist changes in pH.
- CH3COOH and CH3COONa function as acidic buffers, while as NH4CL and NH4OH serve as basic buffers.
- It is noteworthy that our body also possesses its own buffer systems to maintain a stable pH level.
- The buffer capacity of a solution refers to the increase in the overall concentration of particles within the buffer. Consequently, it is associated with a distinct pH value.
- The pH of a buffer solution can be calculated using the Henderson-Hasselbalch equation.

Long Type Questions

1. Give a detailed description of the scope and utilization of biochemistry?
2. Describe in detail the transport system across the cell membrane?
3. Differentiate between;
   a. Active and passive transport.
   b. Exocytosis and endocytosis.
   c. Phagocytosis and pinocytosis
   d. Antiport and symport.

4. Write short notes on:
   a. Diffusion;
   b. Facilitated diffusion;
   c. Osmosis;
   d. Filtration;
5. What is exocytosis? Mention briefly the types of exocytosis?
6. Describe buffers along with their properties and types?
7. Write an elaborated note on Henderson-Hasselbalch Equation?

Fill in the Blanks

1. A movement of specific molecules from a higher concentration gradient to a lower concentration gradient is known as____________
2. ____________ is the movement of a particular molecule against its concentration gradient.
3. Diffusion, osmosis, and filtration are examples of ____________ transport.
4. Cell drinking and cell eating are the common terms for ____________ and ____________
5. The secretion of hormones is an example of ____________
6. CH_3COOH and CH_3COONa function as ____________ buffers.
7. When a solution has a pH value below 7, it is referred to as ____________, while a solution with a pH value above 7 is considered ____________
8. Buffer solutions are composed of a ____________ and its ____________
9. The pH of a buffer solution can be calculated using the____________

2

Carbohydrates

Carbohydrates are the macronutrients consisting of carbon (C), hydrogen (H), and oxygen (O) atoms. In simpler terms, carbohydrates can be defined as polyhydroxy aldehydes or ketones or substances, which yield them on hydrolysis. The general formula of carbohydrates is $C_n(H_2O)_n$, where n is the number of carbon atoms. In nature, carbohydrates are the most abundant biological molecules. In the human diet, carbohydrates are one of the main nutrients that provide energy to the body. The green plants, through the process of photosynthesis, synthesize the carbohydrates.

Classification of carbohydrates

Carbohydrates are also known as saccharides, originating from the Greek word sakcharon, which means sugar. However, not all saccharides have a sweet taste. Carbohydrates can be classified into three distinct groups: monosaccharides, oligosaccharides, and polysaccharides. The number of sugar units determines this classification.

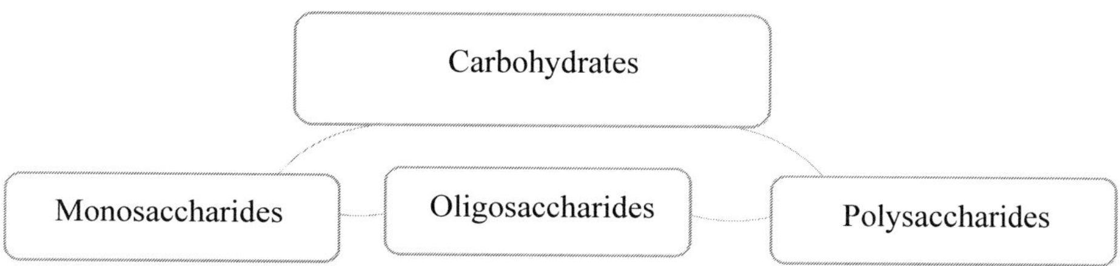

Monosaccharides

Monosaccharides, or simple sugars, are the most basic carbohydrates. They are the monomers that serve as the building blocks for forming polymers or complex carbohydrates. Monosaccharides are the most simple carbohydrates possessing carbonyl groups, i.e., aldehyde (CHO) and ketone (C=O) groups with two or more hydroxyl (OH) groups. Monosaccharide has the generic formula $C_n(H_2O)_{n,}$ where n is the number of carbon atoms. Monosaccharides are sugars that cannot be hydrolyzed further to form simple carbohydrates. These crystalline solids without color dissolve in water but remain insoluble in non-polar solvents.

They can be classed based on the number of carbon atoms, such as triose, tetrose, pentose, hexose, heptoses, and so on, as well as the functional group they possess, such as aldoses (those with aldehyde group) or ketoses (those with ketone group).

Table 1 Classification of monosaccharides

No. of carbon atoms	General term	Empirical formula	Aldose	Ketose
3	Triose	$C_3H_6O_3$	Glyceraldehyde	Dihydroxyacetone
4	Tetrose	$C_4H_8O_4$	Erythrose	Erythrulose
5	Pentose	$C_5H_{10}O_5$	Ribose, Arabinose, Lyxose, and Xylose	Ribulose, Xylulose
6	Hexose	$C_6H_{12}O_6$	Glucose, Mannose, Galactose, Gulose, Idose, Talose, Allose, and Altrose.	Fructose, Sorbose, Psicose and Tagatose
7	Heptose	$C_7H_{14}O_7$	Glucoheptose	Sedoheptulose

Oligosaccharides

Oligosaccharides are complex sugars that, upon hydrolysis, produce two to ten molecules of either distinct or identical monosaccharides. The glycosidic bond joins the monosaccharide units. Oligosaccharides are further grouped as disaccharides, trisaccharides, tetrasaccharides, etc., depending on the number of monosaccharide units in it.

Disaccharides are oligosaccharides that hydrolyze to produce two molecules of monosaccharides. Disaccharides have the general formula $C_n(H_2O)_{n-1}$, including sucrose, lactose, and maltose. Etc. On the other hand, trisaccharides yield three monosaccharide units on hydrolysis, including Raffinose and Rabinose. Moreover, tetrasaccharides are oligosaccharides that yield four monosaccharide units, for example, stachyose ($C_{24}H_{42}O_{21}$).

Polysaccharides

Polysaccharides have more than ten monosaccharide units and can span hundreds of sugar units joined by glycosidic linkage. When broken down by hydrolysis, they give rise to more than ten monosaccharide molecules. They have the general formula $(C_5H_{10}O_5)_n$, where n is the number of carbon atoms and is remarkably large. They are colorless, tasteless, and water-insoluble. Polysaccharides vary in the types of monosaccharide units; they contain their length, the bonds that connect the units, and the amount of branching. These carbohydrates primarily serve structural purposes and store energy.

Homopolysaccharides are composed of monosaccharides of the same type, while heteropolysaccharides are composed of monosaccharides of different types. Examples of homopolysaccharides include starch, glycogen, cellulose, and pectin. Heteropolysaccharides, on the other hand, include hyaluronic acid, heparin, and chondroitin sulfate.

Types of polysaccharides based on their function

Based on their function, polysaccharides are categorized into storage polysaccharides, structural polysaccharides, and complex polysaccharides.

Storage polysaccharides

These are the homopolysaccharides and are the polymers of D-glucose. These are reserved energy sources, with glucose as the body's primary energy source. Thus, they are mainly hydrolyzed into glucose monomers when required. Storage polysaccharides are starch in plants, glycogen in animals, blue-green algae and fungi, and dextran in bacteria and yeast. Other forms of storage polysaccharides are dextrin and inulin.

Structural polysaccharides

Structural polysaccharides are a type of homopolysaccharide formed by the polymerization of D-glucose molecules. They occur in the form of long and strong fibers in the cell walls of plant cells, fungi, and the exoskeleton of arthropods. Cellulose and chitin are the essential structural polysaccharides.

Complex polysaccharides

Complex polysaccharides are composed of different carbohydrates or their derivatives. Like mucopolysaccharides, also known as glycosaminoglycans, are hetero-sugar molecules that form long chains and are present throughout the body. They can be found in mucus and the fluid surrounding the joints. The building blocks of glycosaminoglycans are disaccharide units, which consist of a hexose carbon sugar ring or a hexuronic acid bonded with a hexosamine. One or both monosaccharide units in these chains contain at least one negatively charged sulfate or carboxylate group. The main mucopolysaccharides found in animals are hyaluronic acid, chondroitin sulfate, heparin, and keratin sulfate.

Stereoisomers

Isomers are molecules with the same chemical formula but different structural properties. They are classified as constitutional isomers (with the same chemical formula but different atom connections) and stereoisomers (with the same chemical formula and atom connections but distinct relative 3D orientations). Stereoisomers that differ at all of their stereocenters are mirror images of each other and are known as enantiomers. Stereoisomers that differ at some (but not all) of their stereocenters are known as diastereomers. One of the key features of enantiomers is that they possess a carbon atom with four distinct groups attached to it. This specific carbon atom is referred to as a chiral carbon. When a molecule contains one or more chiral carbons, it is highly probable to exist in the form of two or more stereoisomers. Conversely, with its chiral carbon, glyceraldehyde exists as a pair of enantiomers. D-glyceraldehyde and L-glyceraldehyde are enantiomers, which means they are mirror images of each other. Dihydroxyacetone, lacking a chiral carbon, does not exist as a pair of stereoisomers. Sugars with Fischer projections ending in the same configuration as D-glyceraldehyde are known as D-sugars, whereas sugars derived from L-glyceraldehyde are known as L-sugars.

Epimers

Epimers are the two sugar molecules that differ from each other at a single carbon atom other than functional carbon. The most notable examples of epimers are D-glucose and D-galactose, which vary only at C-4 carbon. Similarly, mannose is a glucose stereoisomer, particularly a C-2 epimer of glucose.

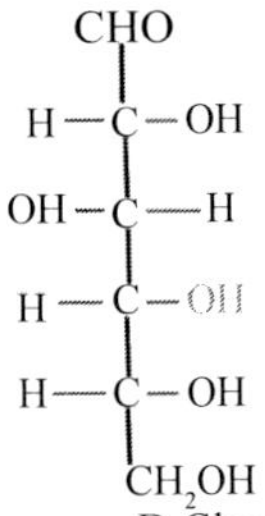

CHO
H—C—OH
OH—C—H
H—C—OH
H—C—OH
CH_2OH
D-Glucose

D-Glucose and D-Galactose are epimers at Carbon 4

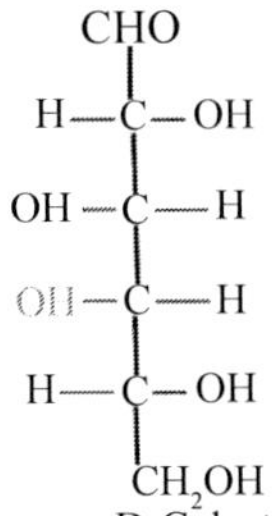

CHO
H—C—OH
OH—C—H
OH—C—H
H—C—OH
CH_2OH
D-Galactose

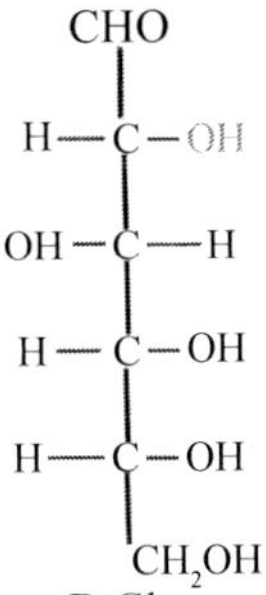

CHO
H—C—OH
OH—C—H
H—C—OH
H—C—OH
CH_2OH
D-Glucose

D-Glucose and D-Mannose are epimers at Carbon 2

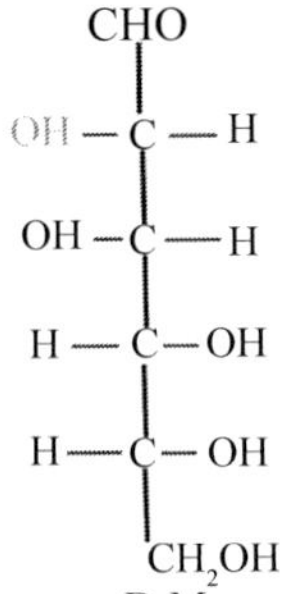

CHO
OH—C—H
OH—C—H
H—C—OH
H—C—OH
CH_2OH
D-Mannose

Anomers

Anomers are a type of isomer found in cyclic monosaccharides or glycosides. They exhibit a difference in the configuration of C-1 for aldoses or C-2 for ketoses. The anomeric carbon (C-1 or C-2), also known as the anomeric center, is the carbon atom that undergoes isomerization in anomers. Depending on the positioning of the anomeric carbon, anomers are categorized as either alpha or beta. The alpha designation is given when the hydroxyl group on the anomeric carbon is oriented downwards in a chair projection (or towards the left in a Fischer projection). In contrast, the beta designation is assigned when the hydroxyl group is directed upwards or towards the right. α-D-glucose and β-D-glucose is an example of anomers because the location of a hydroxyl group at carbon 1(C-1) differs.

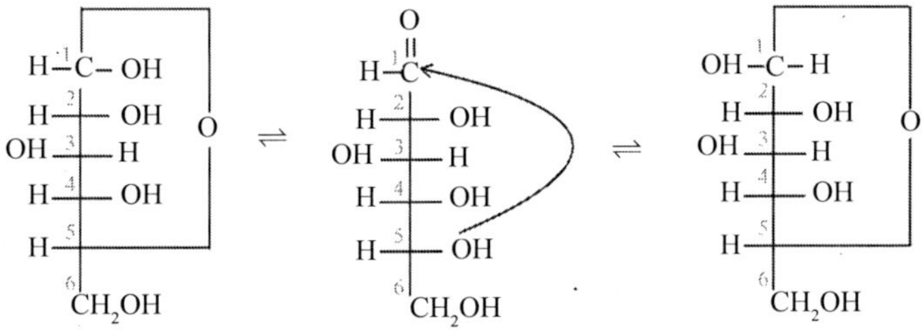

α-D-Glucose ⇌ ⇌ β-D-Glucose

These anomeric carbons are present in reduced form when present in a free state, and if it is linked to another carbon atom, it exists in a non-reducing state. This indicates that all monosaccharides, some disaccharides, and polysaccharides joined by one anomeric carbon are reducing sugars. Similarly, a few disaccharides and polysaccharides joined by all of their anomeric carbons are not reducing sugars.

Mutarotation

Mutarotation refers to the alteration in the optical rotation of a solution caused by a shift in the balance between alpha (α) and beta (β) anomers when dissolved in an aqueous solution. An enzyme known as an isomerase typically catalyzes stereochemistry conversion at a stereocenter. Another way to think about this conversion is to imagine the anomeric carbon in a hemiacetal or hemiketal connection with the oxygen atom in the carbon chain. The word 'Hemi' means 'partial,' which means that an anomeric carbon will be in a bond with the oxygen atom of C-5. A D-glucose sample will eventually settle into an equilibrium that contains around 64% β-D-glucopyranoside and 36% α-D-glucopyranoside. The proportion of each anomer is determined by the sugar molecules' biological properties and environmental factors.

Reducing and non-reducing sugars

Reducing sugars are a class of carbohydrates that possess a free aldehyde or ketone group. These sugars can donate electrons or reduce other molecules. Examples of reducing sugars encompass Glucose, Lactose, galactose, fructose, and Maltose. In contrast, non-reducing sugars do not have free aldehyde or ketone groups, resulting in their inability to reduce other compounds. Sucrose, Trehalose, Raffinose, and Gentianose serve as illustrations of non-reducing sugars.

Important carbohydrates

Glyceraldehyde

Glyceraldehyde is the most basic monosaccharide (Triose). Glyceraldehyde is a molecule found in all living things, including humans, and is an intermediate in the various metabolic processes. Glyceraldehyde is a colorless, sweet substance having a molecular weight of 90.08 g mol^{-1}. Water solubility is 29,200 mgl^{-1}. One asymmetric carbon exists in glyceraldehyde as its only carbon atom is linked to four distinct groups. As a result, there are two potential enantiomers for this molecule, denoted as D- and L-form. When using the Fisher formula, the D-form is shown by placing the OH group to the right of the carbon chain (C-2) when the CH=O group is positioned at the top. In contrast, the L-form displays the OH group on the left side of C-2. Its practicality lies in the fields of nutrition, the production of polyesters and adhesives, as a cellulose modifier, and in the realm of leather tanning. Furthermore, it is a valuable tool in biological research, particularly the 'D' stereoisomer, which acts as a benchmark structure due to its simplicity compared to sugars and amino acids. Scientists utilize the arrangement of D-glyceraldehyde as a reference molecule to analyze the structure and nomenclature (identity) of other molecules.

```
      CH=O                 CH=O
       |                    |
  H — C — OH          OH — C — H
       |                    |
     CH2OH                CH2OH
D-Glyceraldehyde     L-Glyceraldehyde
```

Dihydroxyacetone

Dihydroxyacetone (DHA) is a ketotriose with formula $C_3H_6O_3$ and has a molar mass of 90.078 g mol^{-1}. DHA comprises a ketone group at position 2 and hydroxy groups at locations 1 and 3. It is the most basic member of the ketose class and the parent of the glycerone class. DHA is a hygroscopic white crystalline powder. It has a sweet, refreshing flavor and a distinct odor. It is a sugar molecule with no chiral center. Plant sources like sugar beets and sugar cane often produce it.

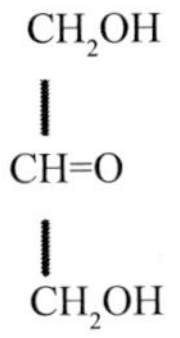

CH_2OH
|
CH=O
|
CH_2OH

Dihydroxyacetone

German scientists discovered DHA as a skin coloring agent in the 1920s. It was discovered that when used in the X-ray process, it caused the skin surface to turn brown when spilled. DHA is most commonly found in sunless tanning products. DHA is sold as an athletic dietary supplement that, when combined with pyruvate, is an orally taken fat burner that enhances lean muscle mass. In addition to these benefits, DHA is known to reduce acne-associated inflammation, accelerate the process of wound healing, reduce hair loss, and boost the growth of hairs.

Erythrose

Erythrose has the chemical formula $C_4H_8O_4$ and is a tetrose saccharide. It belongs to the aldose family because it has one aldehyde group. D-erythrose is the natural isomer; it is a diastereomer of D-threose. Louis Feux Joseph Garot (1798-1869), a renowned French chemist, made a ground-breaking discovery in 1849. He identified erythrose, a compound found in rhubarb and named it after its vibrant red color, attributed to alkali metals' presence. Garot's significant contribution to the chemistry field shed light on erythrose's properties and characteristics. Erythrose is an energy source for oxidative bacteria and plays a role as a plant metabolite.

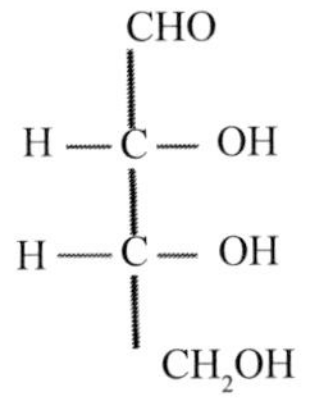

CHO
|
H — C — OH
|
H — C — OH
|
CH_2OH

Erythrose

Erythrulose

D-erythrulose (or erythrulose) is a ketotetrose carbohydrate having the formula $C_4H_8O_4$. Erythrulose is a natural keto-sugar that is transparent to pale-yellowish liquid found in red raspberries. It is made through fermentation using the aerobic bacterium *Glucanobacte*r. This product has no preservatives and is water-soluble. It is used in several self-tanning cosmetics, usually with dihydroxyacetone (DHA). Erythrulose is an effective skincare component for achieving a consistent skin tone.

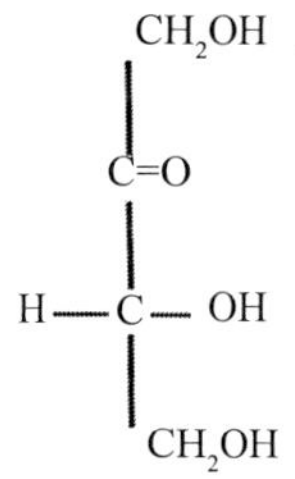

$$
\begin{array}{c}
CH_2OH \\
| \\
C{=}O \\
| \\
H{-}C{-}OH \\
| \\
CH_2OH
\end{array}
$$

Erythrulose

Ribose

Ribose is an aldopentose sugar found in RNA and has the chemical formula of $C_5H_{10}O_5$. In 1891, Emil Fischer and Oskar Piloty were credited with the discovery of ribose. The molar mass of ribose is 150.13 g/mol. Ribonucleoside is formed by the reaction of ribose sugar and a nitrogenous base (adenine guanine, cytosine, and uracil). This ribonucleoside becomes a ribonucleotide when it is linked to a phosphate group. Ribose sugar is a vital component of RNA found within living organisms. RNA plays a crucial role in both encoding and decoding genetic information.

Deoxyribose

Deoxyribose or 2-deoxyribose is also an aldopentose sugar found in DNA that lacks an OH group at position 2 (C-2). The credit for the discovery of deoxyribose goes to Phoebus Levene, who made this significant finding in 1929. Deoxyribose has a hydrogen (H) atom at position 2, whereas ribose has a hydroxyl (OH) group at the same position. Due to this one structural difference, deoxyribose is more stable than ribose sugar. The chemical formula of deoxyribose is $C_5H_{10}O_4$. The molar mass of deoxyribose is 134.13 g/mol.

Its presence is crucial in biological processes. DNA, the fundamental carrier of genetic information in all living organisms, relies on deoxyribose for its structure. The DNA nucleotides consist of various bases, including adenine, thymine, guanine, and cytosine.

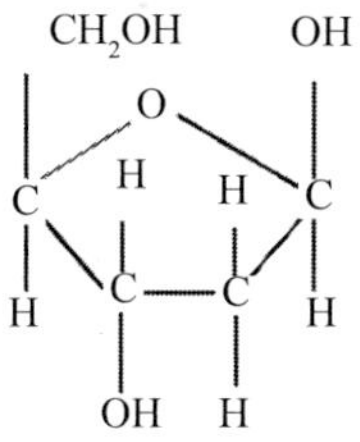

2-Deoxyribose

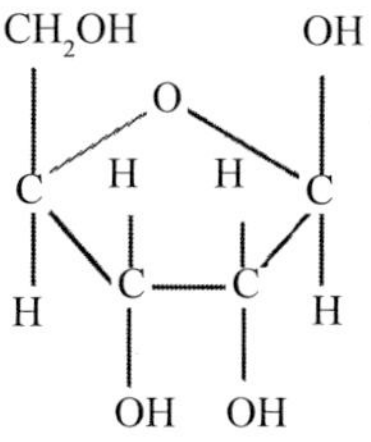

Ribose

Ribulose

Ribulose is classified as a ketopentose, a monosaccharide composed of five carbon atoms and possessing a ketone functional group. Its chemical formula is $C_5H_{10}O_5$. Two enantiomers of ribulose exist: D-ribulose, also referred to as D-erythro-pentulose, and L-ribulose, also known as L-erythro-pentulose. It is important to note that D-ribulose is the diastereomer of D-xylulose. Ribulose sugars are synthesized within the pentose phosphate pathway from arabinose. These sugars play a vital role in synthesizing various bioactive compounds—for example, d-ribulose acts as an intermediate in the fungal pathway for D-arabitol production. Moreover, in green plants, d-ribulose, in the form of 1,5-bisphosphate, combines with carbon dioxide at the onset of the photosynthesis process, serving as a carbon dioxide trap.

D-Ribulose

L-Ribulose

Glucose

Glucose, derived from the Greek word for "sweet," is a widely recognized monosaccharide referred to as dextrose or blood sugar. A German chemist named Andreas Marggraf extracted glucose from raisins in 1747. Jean Baptiste Dumas invented the term glucose in 1838. It consists of six carbon atoms, twelve hydrogen atoms, and six oxygen atoms. It is an aldohexose since it contains one aldehyde group and six carbon atoms. It can exist in two ways: as a ring or as an open-chain structure. It can also exist either in D-form or L-form; however, in nature, it exists predominantly in D-form. It is produced in the kidneys and liver of animals. It can be found in fruits and other plant parts. Glucose is an essential biomolecule that aids in the body's metabolism. This essential sugar is obtained from the food we consume and serves as a vital energy source for our bodies. Termed blood glucose or blood sugar due to its circulation within the bloodstream and subsequent absorption by cells, glucose relies on the hormone insulin to facilitate transportation from the bloodstream to cells for utilization and storage. Individuals with diabetes experience heightened blood glucose levels resulting from insufficient insulin production or an inadequate cell response.

Properties

- Glucose is a white crystalline solid with a melting point of 1460° Celsius.
- When a glucose molecule crystallizes in cold water, glucose monohydrate is generated with the chemical formula $C_6H_{12}O_6.H_2O$ with a melting point of 860°C.
- It is extremely soluble in water and ethanol and entirely insoluble in ether.
- It is approximately three-quarters as sweet as sucrose-containing cane sugar.
- It has optical properties as it rotates plane-polarized light towards the right (dextrorotatory or +), and hence, it commonly exists in nature in d-form.

Different Forms of Glucose Structure

Glucose is an aldohexose that exists in two forms: open chain (acyclic) and ring (cyclic). Glucose can be represented in two ways: Fischer projection, which shows its open chain structure, and Haworth projection, which illustrates its ring structure.

1. Open-chain formula

Baeyer first proposed the open-chain structure of glucose. Glucose's open-chain structure has the following properties:

- There are six carbon atoms.
- There are five hydroxyl groups.
- One aldehyde group on carbon number 1 (C-1)
- Each of the remaining five carbon atoms has a hydroxyl group.
- The open-chain isomer of glucose is often found in small and intermittent levels in living organisms. D-glucose is the most common isomer, and L-glucose is rare.

CHO	CHO	CHO
H—C—OH	H—C—OH	H—C—OH
OH—C—H	H—C—OH	OH—C—H
H—C—OH	H—C—OH	OH—C—H
H—C—OH	H—C—OH	H—C—OH
CH_2OH	CH_2OH	CH_2OH
D-Glucose	Glucose	L-Glucose

2. Configuration

Glucose, also called D-(+)glucose, is characterized by its configuration denoted by the letter 'D' and its dextrorotatory nature indicated by the symbol '(+).' The configuration of D/L-glucose provides insight into the structure of the glucose compound and can be elucidated through the following points. When the letters 'D' or 'L' precede a glucose compound, it signifies the relative configuration of a specific stereoisomer of the glucose compound to the configuration of other compounds. For instance, the D-isomer of glyceraldehyde possesses a 'D' configuration, which implies that the OH group is positioned on the right-hand side of its structure. On the other hand, compounds that can be linked to the 'L' isomer of glyceraldehyde are described as having L-configuration, where the OH group is situated on the left-hand side. Compounds that rotate the plane of polarization towards the right are dextrorotatory (d or +), and those that rotate towards the left are laevorotatory (l or -).

Glucose is a widely prevalent and significant carbohydrate, naturally existing as the D-(+)- isomer. Conversely, L-glucose is artificially synthesized. Compared to other D-L configurations, D-glucose and D-mannose are referred to as epimers, differing solely at the C-2 position. Similarly, D-glucose and D-galactose are also recognized as epimers, differing in C-4 configuration.

3. Cyclic structure

Upon careful examination, Fischer discovered that the chemical behavior of glucose could not be fully elucidated by its open-chain pentahydroxy aldehyde structure. Glucose behaves differently from typical aldehydes. It does not form a crystalline bisulfate compound or show a positive reaction in Schiff's test. Additionally, glucose derivatives such as pentaacetate and pentamethyl ether do not undergo oxidation when treated with Tollen's reagent or Fehling's solution, suggesting the absence of the aldehyde group. The French chemist Tarnet identified two crystalline forms of glucose: alpha-glucose and beta-glucose. It was observed that alpha-glucose exhibited a specific rotation of +112°, whereas beta-glucose displayed a specific rotation of +19°. As time progressed, the optical rotation of both forms underwent gradual changes until it eventually stabilized at a constant value of +53°. This intriguing phenomenon, known as mutarotation, was explained by postulating that alpha and beta glucose were cyclic hemiacetal forms of glucose, capable of interconversion through the open-chain form.

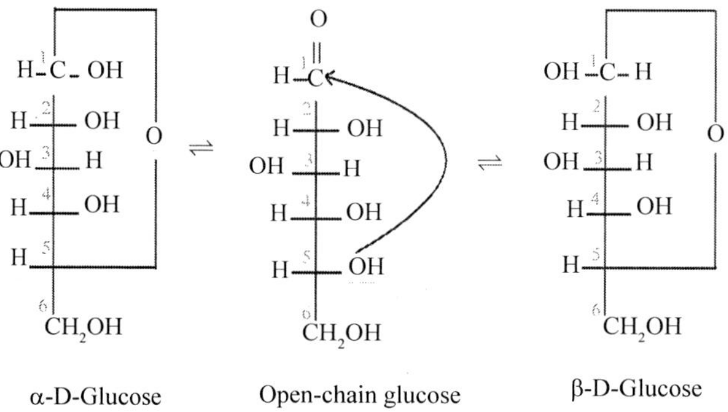

The confirmation of a cyclic structure is achieved through glycoside formation. When glucose is treated with methanol in the presence of dry HCl, it yields two isomeric glycosides or acetals. These glycosides, namely methyl-α-D-glucose and methyl-β-glucoside, are isolated in a crystalline state. Although they possess optical activity, they do not exhibit any reactions associated with the free aldehyde group.

4. Haworth representation

Haworth held the view that these cyclic configurations were awkward. Consequently, he devised the hexagonal representations, resembling the heterocyclic pyran encompassing five carbon atoms and one oxygen atom within the ring. He introduced the designations α-D-glucopyranose and β-D-glucopyranose for the hexagonal structures corresponding to α-D-glucose and β-D-glucose, respectively. It is important to note that in the Haworth formula, all the hydroxyl (OH) groups on the right side of the Fischer formula are oriented below the plane of the ring, while those on the left side are positioned above the plane of the ring. The terminal CH_2OH group projects its hydroxyl group above the plane of the ring in a glucose molecule.

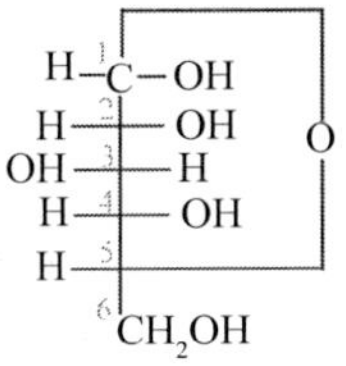

α-D-Glucose
(Fischer)

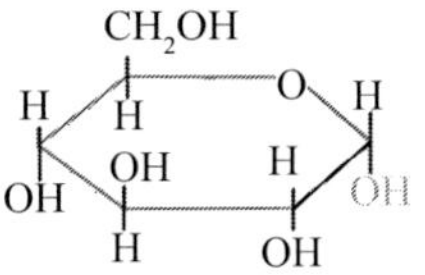

α-D-Glucopyranose
(Haworth)

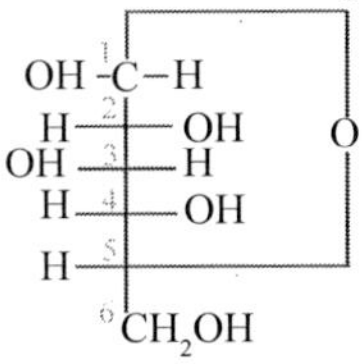

β-D-Glucose
(Fischer)

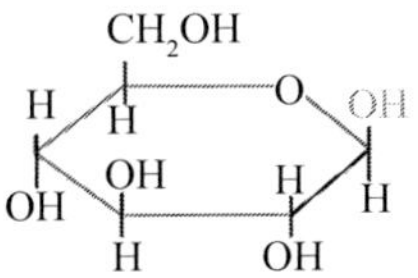

β-D-Glucopyranose
(Haworth)

Uses of Glucose

- It is used to treat hypoglycemia (low blood sugar).
- Because it provides carbohydrate calories, it is administered to individuals who cannot eat.
- It is used to treat high potassium levels in the blood (hyperkalemia).

Fructose

Fructose, also known as fruit sugar or levulose, is a naturally occurring ketohexose. This simple ketonic monosaccharide sugar can be found in various plants, flowers, and fruits. Alongside glucose and galactose, fructose is one of the three monosaccharides the body absorbs as part of the diet. Due to its simplicity, fructose is easily digested compared to other sugars. The credit for the discovery of fruit sugar goes to Augustin-Pierre Debrunfaut, a French scientist. It is commonly present in fruits like mango, litchi, cherry, and guava, as well as vegetables such as carrot, radish, beetroot, and sugarcane. This sugar is commercially generated from corn, sugar cane, and sugar beets. However, if taken in excess, it can lead to obesity and insulin resistance, to mention a few side effects.

Fructose Structure

Crystalline fructose displays a ring structure, specifically a polyhydroxy ketone with 6 carbons. The presence of hemiketal and internal hydrogen bonding contributes to its stability. In its crystalline form, it is referred to as D-fructopyranose. However, when dissolved in water, fructose undergoes an equilibrium combination where 70% exists as fructopyranose, 22% as fructofuranose, and the remaining 7% comprises three other forms, including its acyclic structures.

Cyclic Form

The molecular structure of fructose consists of six carbon atoms, twelve hydrogen atoms, and six oxygen atoms. Its unique characteristic lies in the presence of a ketone functional group at carbon number 2, making it a ketohexose. Fructose and glucose share the same chemical formula, $C_6H_{12}O_6$.

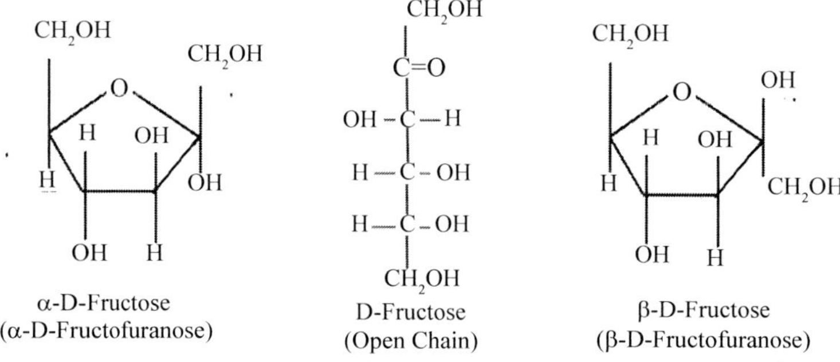

α-D-Fructose (α-D-Fructofuranose)

D-Fructose (Open Chain)

β-D-Fructose (β-D-Fructofuranose)

Haworth projection of the furanose form

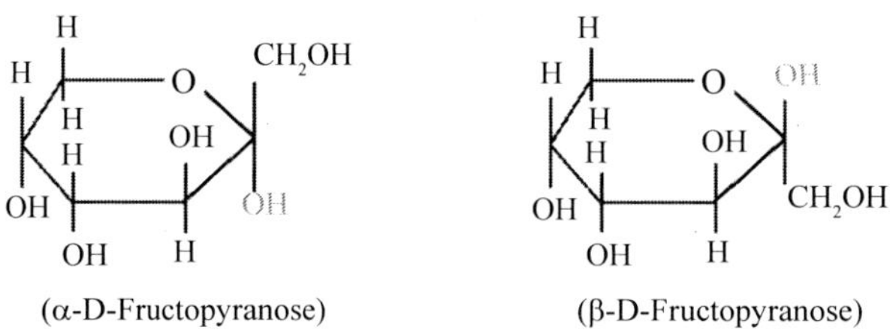

(α-D-Fructopyranose)

(β-D-Fructopyranose)

Haworth projection of the pyranose form

D-Fructose Fischer projection

α-D-Fructose

β-D-Fructose

Hemiacetal ring closure

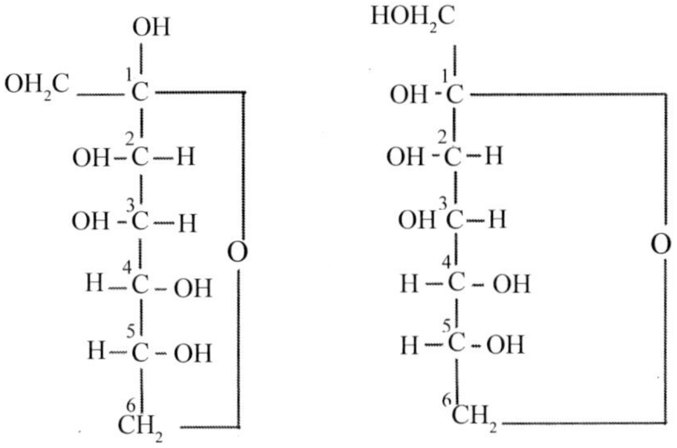

```
          OH                    HOH2C
          |                       |
OH2C ___ 1C ------+         OH - 1C -----------+
          |       |               |            |
     OH - 2C - H  |         OH - 2C - H        |
          |       |               |            |
     OH - 3C - H  |         OH   3C - H        O
          |       O               |            |
      H - 4C - OH |          H - 4C - OH       |
          |       |               |            |
      H - 5C - OH |          H - 5C - OH       |
          |       |               |            |
         6CH2 ----+              6CH2 ---------+
```

α-D-Fructose β-D-Fructose

Hemiacetal ring closure

Physical properties

Fructose is an essential ketohexose that is produced through the process of hydrolysis of the disaccharide sugar (sucrose). The following are the physical properties of fructose:

- It has a molecular mass of 180.156 $g.mol^{-1}$.
- It has a melting point of 103° C.
- It is the most water-soluble sugar.
- It has a sweet flavor. This is why it is used as a sweetening agent in beverages and meals. It is the cheapest and tastiest naturally occurring carbohydrate. Its relative sweetness reduces as the temperature rises.
- At normal temperature, it is a white crystalline solid.
- It's a flavorless sugar.
- It possesses a high hygroscopicity, or the ability to absorb moisture from its surroundings. Compared to other sugars, it absorbs moisture quickly and releases it slowly.
- It is a good humectant, which means it can hold moisture.

Chemical Properties

It is extremely important to understand all of the chemical properties of fructose because it is widely used in manufacturing food and is part of our daily diet. The following are the chemical characteristics of fructose:

- Yeast and bacteria can ferment fructose anaerobically. Yeast breaks down sugar into ethanol and carbon dioxide.
- It can contribute to the generation of mutagenic compounds.
- Fructose on dehydration yields hydroxymethylfurfural ("HMF").

Fructose Applications

Fructose has a wide range of applications. A few examples are provided below.

- Because it may be transformed into hydroxymethylfurfural ("HMF"), it can be utilized in diesel fuel additives as well as diesel fuel.
- Fructose is widely used as a sweetener on a large scale. It is used as a sweetener in a variety of cuisines. It is a low-calorie sweetener, which increases its usefulness as a sweetener.
- It is also utilized in polymers.
- It is used as a newborn feeding formula.
- It is used to extend the life of goods such as nutrition bars and cookies.
- Fructose is used to make corn syrup.
- It is used as a diabetic meal.
- It has a low glycemic index, resulting in moderate insulin release into the bloodstream compared to glucose and fructose.
- It is found in various beverages, baked goods, and other products.

Galactose

Galactose (milk sugar), often called Gal, is a monosaccharide sugar similar to glucose and 65 percent sweeter than sucrose. Charles Weissman proposed the term galactose, which is taken from the Greek word galaktos, which refers to milk, and ose which stands for sugars. Galactose's chemical formula is $C_6H_{12}O_6$. It is a C-4 epimer of glucose and is an aldohexose. Galactose is a simple sugar that occurs in lactose in the D-form.

Galactan, a naturally occurring polymeric carbohydrate group, is composed of galactose and is commonly found in hemicellulose. It serves as the fundamental structure of galactans, which are polymeric forms of galactose. Dairy products, avocados, sugar beets, and various gums and mucilages are some of the sources where galactose can be found. Additionally, galactose is naturally produced within the body and can be present as glycolipids and glycoproteins in numerous tissues. Moreover, it is worth noting that galactose is a by-product of the third-generation ethanol manufacturing process derived from macroalgae.

Properties

- Galactose has the chemical formula $C_6H_{12}O_6$.
- It is a white solid without an odor.
- Galactose has a molecular mass of 180.156 g/mol.
- Galactose has a density of 1.5 $g.cm^{-3}$.
- Galactose has a melting point of 168-170° C.

Galactose Structure

The structure of galactose is asymmetric, resulting in different shapes on each side. It is also a chiral compound, with the left-handed form referred to as D-galactose and the right-handed form known as L-galactose. D-galactose and L-galactose are the two enantiomers of galactose, an aldohexose. According to the Fischer-Rosanoff convention or D-L system, D-galactose naturally occurs in higher organisms. Therefore, the term galactose is commonly used to refer to D-galactose. Galactose can exist in both linear or open-chain form and cyclic form. The open-chain form, represented by a Fischer projection, contains an aldehyde or carbonyl group at one end of the chain. This carbonyl group allows galactose to be classified as a reducing sugar, similar to other monosaccharides and disaccharides like lactose and maltose. Galactose has two structural forms - open chain and cyclic structure. It is considered a C-4 epimer of glucose, meaning that the hydroxyl group on the fourth carbon has a different position compared to glucose. Galactose has four cyclic isomers, with two isomers having a pyranose ring (a six-membered ring) and the other two having a furanose ring (a five-membered ring). In solution, the open-chain configuration is not thermodynamically stable and exists in minimal quantities. The equilibrium between the pyranose and furanose forms is shifted towards the pyranose form due to its greater stability compared to the furanose form. Among mammals, only the pyranose form is observed in the cyclic structure. In contrast, the furanose form is prevalent in various lower organisms, including bacteria, green algae, fungi, protozoa (such as Trypanosomatids), sponges, and starfish. The furanose form also plays a role in pathogenic species, potentially contributing to their survival, and targeting its biosynthetic pathway could be a promising approach for developing new drugs to treat infections caused by these pathogens.

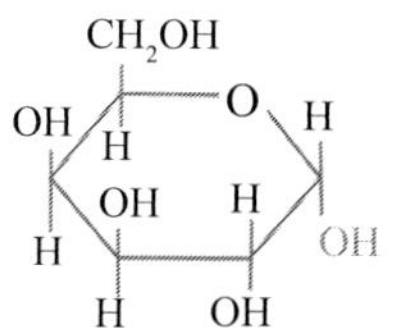

(α-D-Galactopyranose)

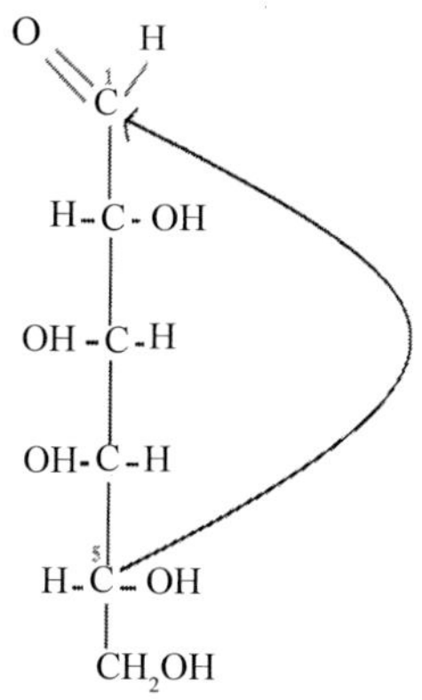

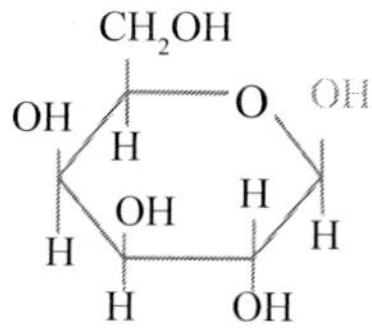

β-D-Galactopyranose

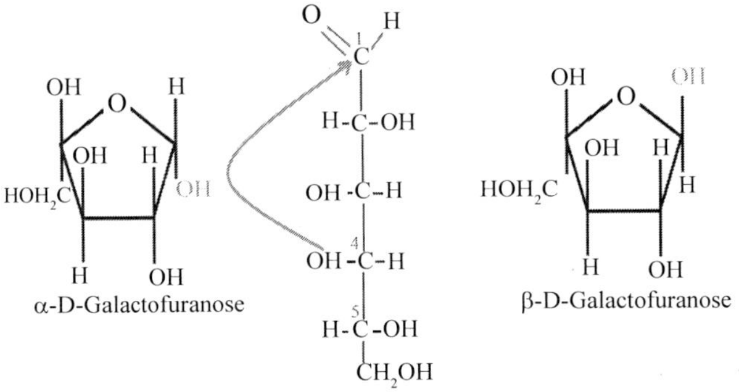

α-D-Galactofuranose

β-D-Galactofuranose

Galactosemia is a genetic condition that occurs when the body cannot properly break down galactose due to a mutation in one of the enzymes involved in the Leloir pathway. This pathway is responsible for metabolizing galactose in the body. When there is a mutation in one of these enzymes, galactose cannot be processed correctly, leading to a build-up of galactose in the body. Galactosemia is an inherited condition that is passed down from parents to their children through genes. Individuals with galactosemia must strictly avoid consuming foods that contain galactose, such as milk and dairy products, as it can lead to serious health complications. Early diagnosis and management of galactosemia are crucial to prevent long-term complications and ensure the affected individuals can lead healthy lives. As a result, galactosemics are toxic to even trace levels of glucose.

Mannose

Mannose is a simple hexose sugar found naturally in plants such as cranberries. Mannose is not considered an essential food; it can be generated or converted into glucose in the human body. Mannose can be produced by oxidizing mannitol. Mannose contains 2-5 kcal/g. It is eliminated in part through the urine. Mannose is essential in human metabolism and post-translational modification, particularly in the glycosylation of specific proteins. Mannose (D-mannose) is a dietary supplement used to prevent recurring urinary tract infections (UTIs).

Structure

Mannose is typically found in two forms of rings: pyranose (six-membered) and furanose (five-membered). Each ring closure can have an alpha (α) or beta (β) configuration at the anomeric position. The chemical rapidly isomerizes between these four forms. When dissolved in water, mannose forms a mixture comprising α-D-mannopyranose (63.7%) and β-D-mannopyranose (35.5%) isomers. Similarly, the mannose also forms two five-membered (furanose) ring structures as α-D-Mannofuranose (0.6%) and β-D-Mannofuranose (0.2%).

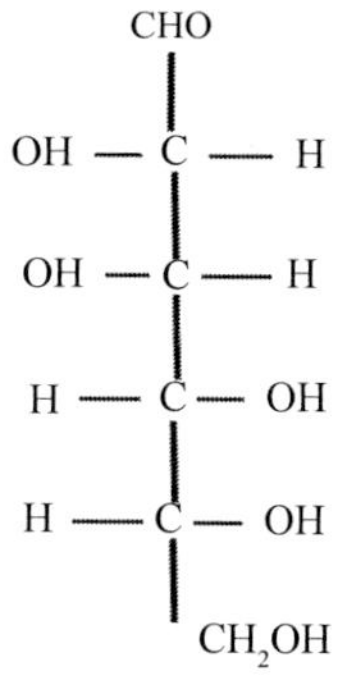

CHO

OH — C — H

OH — C — H

H — C — OH

H — C — OH

CH_2OH

D-Mannose

D-Mannose isomers (Haworth Projections)

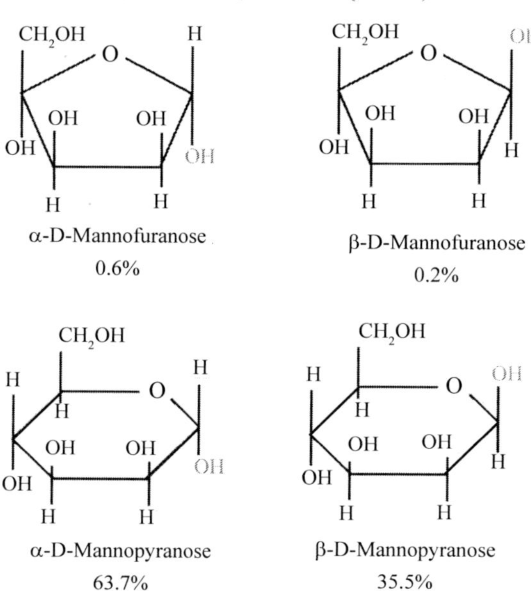

α-D-Mannofuranose 0.6%

β-D-Mannofuranose 0.2%

α-D-Mannopyranose 63.7%

β-D-Mannopyranose 35.5%

Maltose

Maltose also known as Cextromaltose, Maltobiose, Maltodiose, and D-Maltose. Maltose is a disaccharide composed of two D-glucose units joined by an α (1→4) glycosidic bond and having the chemical formula $C_{12}H_{22}O_{11}$. Augustin-Pierre Dubrunfaut, a French chemist, 'discovered' maltose. His discovery was not widely acknowledged until Cornelius O'Sullivan, an Irish chemist and brewer, validated it in 1872. The word maltose was created from malt, and the suffix '-ose,' which is common in sugar names—Maltose production and use in China date back to the Shang era. Maltose has been used and recorded since the reign of Emperor Jimmu.

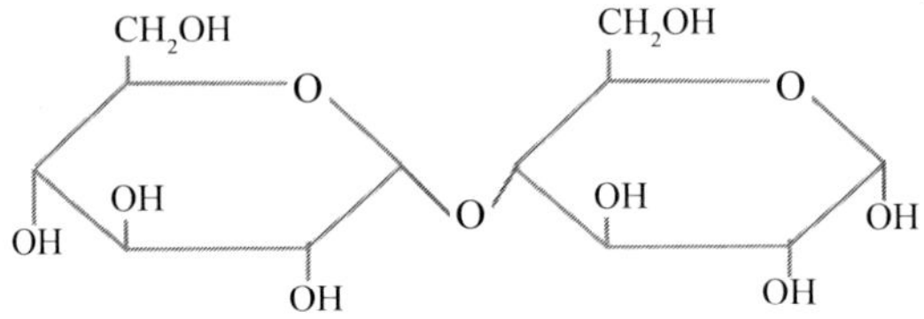

Maltose

Maltose is a disaccharide in which the first and fourth carbons of the glucose units (i.e., the first carbon of one glucose unit and the fourth carbon of another glucose unit) are linked through a glycosidic bond. At the first carbon of the second glucose molecule, a free aldehyde group is present, and due to this fact, maltose is a reducing sugar. In mammals' maltose is produced from the digestion of starch by the action of the β-amylase enzyme, and then the maltase enzyme converts maltose to glucose units. In an aqueous solution, maltose exhibits mutarotation, resulting in the equilibrium between two different forms. The identification of maltose can be conveniently accomplished through the application of the Woehlk test or Fearon's test using methylamine. Maltose is between 30-60% sweeter than sugar in concentration.

- Maltose can be found as a white powder or crystals.
- Maltose produces carbon dioxide, water, and sulfur dioxide when it reacts with sulfuric acid:

  $12CO_2 + 35H_2O + 24SO_2 = C_{12}H_{22}O_{11} + 24H_2SO_4$
- The hydrolysis of maltose produces ethanol and carbon dioxide:

  $4C_2H_5OH + 4CO_2 = C_{12}H_{22}O11 + H_2O$
- Maltase enzyme works as a catalyst, resulting in maltose breakdown into glucose:

Uses of Maltose

- In the form of corn syrup, maltose serves as a low-cost sugar source.
- Maltose is used in the malting process of barley to make beer; the malting process adds sweetness to the beverage.
- Because of its caramel-like flavor, it is used in bakeries, soft drinks, candies, alcoholic beverages, infant food, and sugar-free items.
- Maltose is synthesized during seed germination by an enzyme called amylase, which hydrolyzes starch to disaccharide for the emerging plants.
- During digestion, amylase partially turns starch into maltose, and maltase released by the intestine converts maltose into glucose, which is either used by the body or stored in the liver as glycogen.

Isomaltose

Isomaltose is a disaccharide with the molecular formula $C_{12}H_{22}O_{11}$ that comprises two glucose units. This substance has a molar mass of 342.30 g mol^{-1}. Its melting point ranges from 98 to 160° Celsius. It is soluble in water and is a sugar that ferments. Isomaltose is a kind of maltose isomer, and the two are distinguished by the glycosidic linkage that unites two glucose molecules. In isomaltose, the linkage is α (1→6) glycosidic bond, whereas in maltose, it is α (1→4). Isomaltose, like maltose, is a reducing sugar. The molecule can adopt an open-chain configuration by selectively involving one of the two anomeric carbons in the glycosidic bond. In this configuration, the functional group assumes the role of a reducing agent. The digestion of carbohydrates such as bread, rice, and potatoes can also yield Isomaltose. The hydrolysis of α (1→6) linkages is more difficult than that of α (1→4) bonds. A glucosidase enzyme called sucrase-isomaltase assists in the hydrolysis of α (1→6) linkages. The enzyme is situated on the brush edge of the small intestine. In addition to isomaltose, it aids in the digestion of sucrose and starch. As a result, a deficiency in this enzyme may result in indigestion and osmotic diarrhoea.

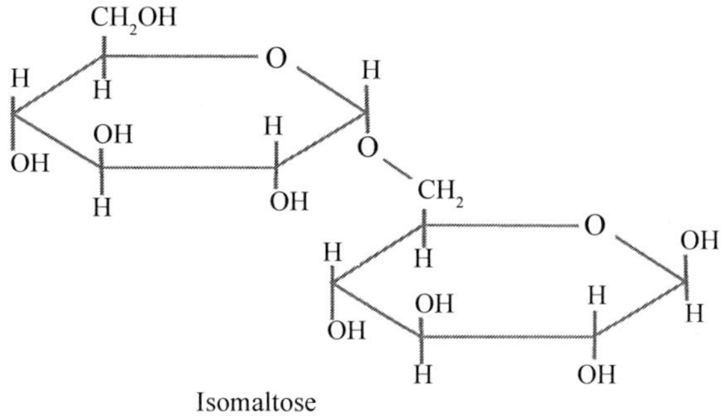

Isomaltose

Importance

Isomaltose is a disaccharide derived from glucose that may be digested to produce chemical energy. Therefore, isomaltose may be used as a source of energy as it is formed after the digestion of starch or meals containing isomaltose. The majority of dietary isomaltose, however, is synthesized rather than obtained naturally. It is instead achieved with artificial preparations. For example, it is generated artificially from high maltose syrup with the enzyme transglucosidase. Another way is to break down sucrose using the bacterial enzyme sucrose isomerase.

Isomaltose is an artificial sweetener used in confections, candies, and canned goods. In addition to the food industry, artificial isomaltose is used in animal feed, pharmaceuticals, and cosmetics.

Isomaltose Deficiency

Sucrase-isomaltase (SI) insufficiency is a rare metabolic condition characterized by an inability to digest particular sugars due to deficiency or absence of enzymes, i.e., sucrose-isomaltase. Without these enzymes, isomaltulose, sucrose, and isomaltose are unabsorbed and act as osmotic laxatives. Common symptoms include diarrhoea, vomiting, gas, and stomach distention. The condition is connected to mutations in the SI gene, as this gene codes for the sucrase-isomaltase enzyme.

This gene's mutation may have affected the enzyme's structure, function, or production. The disorder is inherited in an autosomal recessive manner. The sucrase-isomaltase deficit is treated by a diet adjustment that reduces the amount of starch or sucrose ingested. Genetic testing is also an alternative to screening.

Lactose

Lactose, commonly known as milk sugar, is a disaccharide found in the milk of various mammals, such as humans and cows. Its molecular formula, $C_{12}H_{22}O_{11}$, indicates that it comprises 12 carbon atoms, 22 hydrogen atoms, and 11 oxygen atoms. In 1843, lactose was given its name by the French chemist Jean Baptiste Andre Dumas (1800-1884). Subsequently, in 1856, Pasteur assigned the name "lactose" to galactose. However, in 1860, Marcellin Berthelot renamed it "galactose" and reassigned the term "lactose" to its current definition, referring to the specific substance known as lactose today. Lactose is a disaccharide that combines D-galactose and D-glucose, linked through the β (1→4) glycosidic bond. Scientifically, it is referred to as β-D-galactopyranosyl-(1→4) D-glucose. The glucose ring can exist in either the pyranose or the furanose form, while galactose can only exist in the pyranose form. Therefore, two specific forms, i.e., α-lactose and β-lactose, describe the anomeric form of the glucopyranose ring. Consequently, a lactose solution contains both α and β anomers at the "reducing end" of the disaccharide. Lactose, classified as an aglycone hemiacetal, exhibits mutarotation, classifying it as a reducing sugar. The opening of the lactose ring forms a free aldehyde group that can react with Benedict's solution.

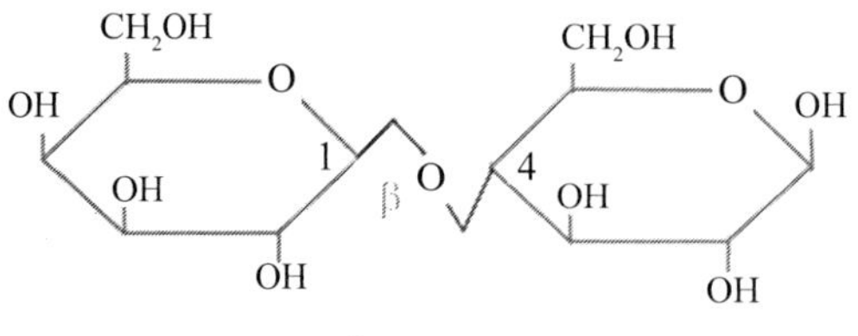

Lactose

Woehlk and Fearon's tests are chemical reactions used to detect the presence of lactose. These tests are commonly employed in educational settings to demonstrate the lactose content in various dairy products, including whole milk, lactose-free milk, yogurt, buttermilk, coffee creamer, sour cream, and kefir. Lactose, a white solid, possesses a mildly sweet taste and can dissolve in water. Additionally, it is a solid that does not absorb environmental moisture. With a molar mass of 342.29 g/mol and a 1.525 g/cm^3 density, lactose undergoes enzymatic hydrolysis in the intestine by the "lactase" enzyme, breaking it down into glucose and galactose. In alkaline solutions, lactose can be isomerized to lactulose, and under catalytic hydrogenation, it can be converted to the polyhydric alcohol lactitol. Lactulose, a by-product of lactose, is a commercial medication for relieving constipation.

Uses

1. Lactose serves multiple purposes, including being utilized as a nutrient in the formulation of modified milk and as an additive in infant food to replicate the composition of human milk.
2. Besides aiding in calcium absorption, lactose plays a crucial role in facilitating the body's uptake of minerals such as magnesium, copper, and zinc.
3. Due to its advantageous physical and functional properties, lactose is incorporated into pharmaceutical products as an excipient and diluent for tablets and capsules.
4. The food industry harnesses lactose as a flavoring agent, enhancing the taste of numerous food products.
5. Lactose finds application in the production of baked goods and is also utilized in fermentation processes to create cheese, yogurt, and sour milk.
6. Furthermore, lactose serves as a chromatographic adsorbent in the field of analytical chemistry

Sucrose

Commonly known as table sugar or cane sugar, Sucrose is a disaccharide composed of α-D-glucose and β-D-fructose, linked together by a (α1→β2) glycosidic bond. The chemical formula of sucrose is $C_{12}H_{22}O_{11.}$ William Miller, an English chemist, coined the term "sucrose" in 1857. The sucrose molecule lacks any anomeric hydroxyl groups. Hence, sucrose is categorized as a non-reducing sugar since it does not function as a reducing agent.

Sucrose is predominantly found in various plants, particularly in their roots, fruits, and nectars, within the realm of nature. It serves as a means to store energy primarily derived from photosynthesis. Numerous mammals, birds, insects, and bacteria accumulate and rely on sucrose from plants as their primary source of sustenance. From a human consumption standpoint, honeybees play a crucial role as they gather sucrose and produce honey, a significant foodstuff worldwide.

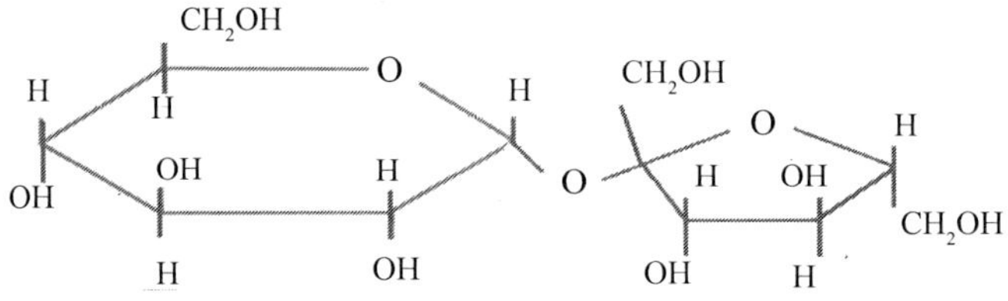

Sucrose

Honey predominantly contains fructose and glucose as its main carbohydrate components, while sucrose is found in minimal quantities. During the ripening process, the sucrose content in fruits generally increases significantly. However, certain fruits, such as grapes, cherries, blueberries, blackberries, figs, pomegranates, tomatoes, avocados, lemons, and limes, contain negligible amounts of sucrose.

Physical properties

Sucrose has a monoclinic crystal structure and is highly soluble in water. Upon exposure to elevated temperatures (above 186° C), this substance undergoes decomposition, resulting in the formation of caramel. At 20° C, its solubility in water is measured to be 203.9g/100mL. Its defining characteristic is its sweet taste.

Chemical properties

- The combustion of sucrose yields carbon dioxide and water as its by-products.
- Hydrochloric acid, carbon dioxide, and water are produced when this compound is combined with chloric acid.

  $$C_{12}H_{22}O_{11} + 8HClO_3 \rightarrow 8HCl + 11H_2O + 12CO_2$$

- During hydrolysis, the glycosidic linkage that joins the two carbohydrates within a C12H22O11 molecule is cleaved, leading to the production of glucose and fructose.
- By utilizing H_2SO_4 as a catalyst, sucrose can be dehydrated, forming a carbon-rich black solid.

  $$C_{12}H_{22}O_{11} + H_2SO_4 \rightarrow 11H_2O + 12C \text{ (carbon-rich solid)} + \text{heat}$$

Uses of sucrose

Below are some of the significant applications of sucrose

1. Sucrose plays a crucial role as a key ingredient in soft drinks and various other beverages. 2. This compound is extensively used in producing numerous pharmaceutical products.
3. It serves as a chemical precursor for the production of emulsifying agents and detergents.
4. Moreover, it acts as a thickening agent and stabilizer in food products.
5. The inclusion of this compound helps in prolonging the shelf lives of various food products, such as jams and jellies.
6. When used in baking, sucrose contributes to the desirable brown color of the baked goods.
7. Furthermore, this compound exhibits antioxidant properties, effectively inhibiting oxidation.
8. Sucrose is widely employed as a food preservative, ensuring the preservation of food items.

Cellobiose

Cellobiose is classified as a reducing sugar and has the chemical formula $(C_6H_7(OH)_4O)_2O$. Combining two β-D-glucose molecules forms its chemical structure through a β (1→4) glycosidic bond.

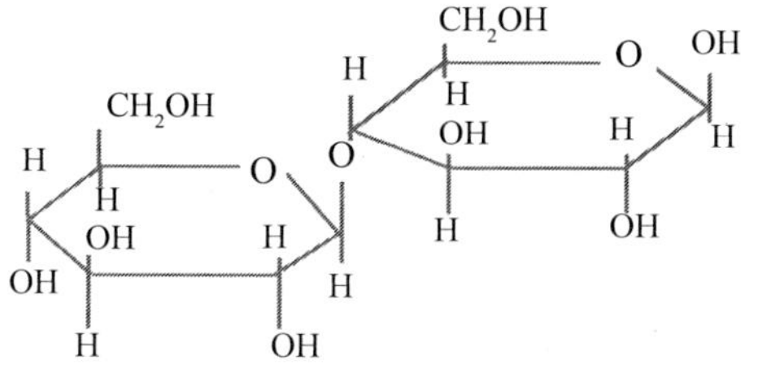

Cellobiose

By undergoing enzymatic or acidic hydrolysis, cellobiose can be broken down into glucose. With eight free alcohol (OH) groups, one acetal linkage, and one hemiacetal linkage, cellobiose forms strong inter and intramolecular hydrogen bonds, resulting in its solid white appearance. Cellulose and cellulose-rich materials such as cotton, jute, or paper can be hydrolyzed to produce cellobiose. When cellulose is treated with acetic anhydride and sulfuric acid, it forms cellobiose octoacetate. This derivative loses its ability to donate hydrogen bonds (although it can still accept them) and becomes soluble in nonpolar organic solvents. Additionally, cellobiose is used as a diagnostic carbohydrate for Crohn's disease and malabsorption syndrome.

Starch

The primary dietary source of carbohydrates is starch $(C_6H_{10}O_5)_{n,}$ where n represents the number of α-D-glucose units. It is usually found in cereals, potatoes, legumes, and other plants. When combined with an iodine solution, natural starch becomes blue and is insoluble in water. Starch is a mixture of two compounds, amylose and amylopectin.

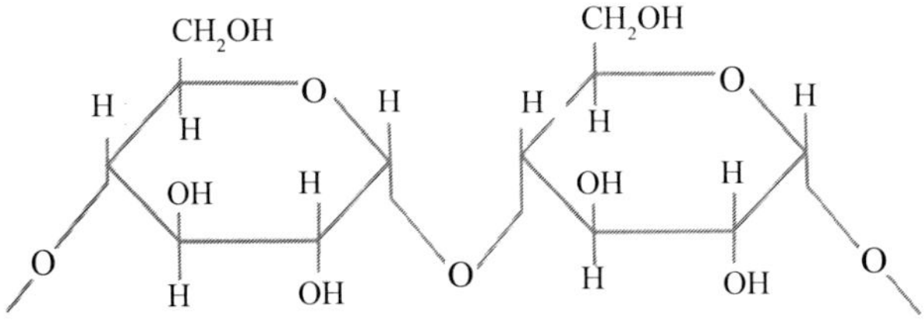

Strach

Amylose

Amylose, a starch component, is a polymer composed of D-glucose units linked together by α (1→4) glycosidic bonds. This linear chain consists of numerous glucose molecules, ranging from hundreds to thousands in number. With its water-soluble nature, amylose constitutes around 20-25% of starch content. It can adopt three forms: one amorphous structure and two helical conformations. In plants, amylose functions as a storage unit for sugar and energy. This polysaccharide has the remarkable ability to swell in water and is resistant to the action of pancreatic α-amylase. Additionally, enzymes present in the colonic microflora can break it down. The utilization of amylose can play a pivotal role in supporting optimal digestion, fortifying the immune system, facilitating weight loss, managing diabetes symptoms, and reducing the likelihood of heart disease. Moreover, it finds applications as a binding agent in culinary preparations, as well as in the production of plastics and paper.

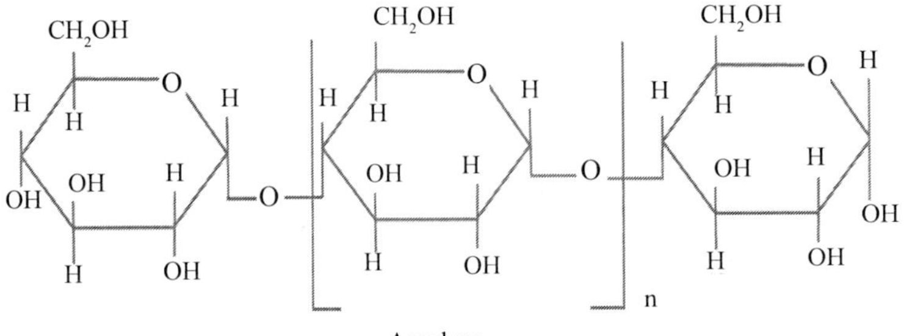

Amylose

Amylopectin

Amylopectin makes 80 to 85 % of starch. The branched chain has 24 to 30 α-D-glucose residues that make up this structure. In the main chain, α (1→4) links hold the glucose residues together, and at the branch points, α (1→6) glycosidic bonds.

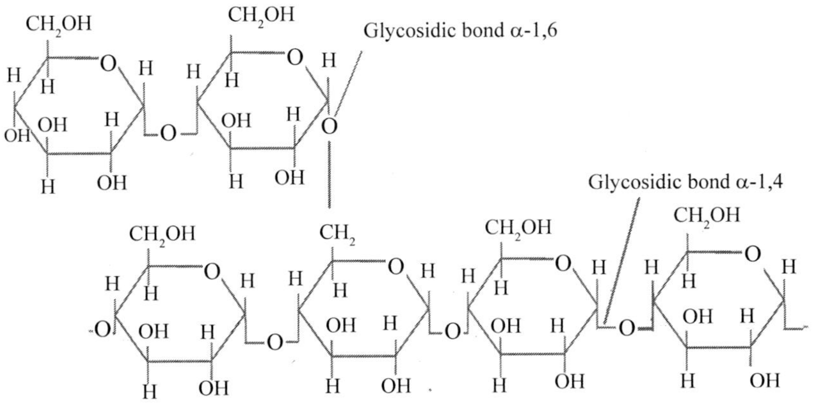

Amylopectin

Amylopectin is a fraction that is not soluble in water. It is considered a reducing sugar due to the presence of reducing ends. The length and branching points of amylopectin chains determine their classification into three types:

1. **A chains**: These are the shortest chains with no branch points.
2. **B chains:** They are branched by either A chains or other B chains.
3. **C chains:** These chains carry other B chains and contain the only reducing terminal residue.

Amylopectin finds various applications, primarily as a thickening agent or stabilizer.

Glycogen

The counterpart of starch in the animal body is glycogen, which occurs in significant amounts in the liver and muscles. Glycogen is a readily mobilized storage form of glucose. It is stored in the cytoplasm in the form of hydrated granules. It is a branched polysaccharide composed entirely of α-D-glucose units connected by α (1→4) glycosidic bond in the main chain. In contrast, the linkages at the branch point are between carbon atoms 1 and 6, connected by the α (1→6) bond. Glycogen is non-reducing and gives a red color with iodine.

Glycogen

Functions of glycogen

1. The liver glycogen acts as a storage form of glucose, which hepatocytes release to regulate the blood glucose levels within the normal range. Although about 40 kcal of glucose is available in body fluids, the hepatic glycogen reserves can supply approximately 600 kcal after a fasting period.
2. The glucose obtained from glycogen reserves remains confined within the cells of skeletal and cardiac muscles, serving as a fundamental energy source for muscle activity. In the brain, astrocytes contain a limited amount of glycogen. This glycogen is accumulated during sleep and subsequently mobilized during periods of wakefulness and activity.
3. The presence of glycogen reserves in the brain also provides a certain level of protection against hypoglycemia or low blood sugar.
4. Furthermore, glycogen is specialized in fetal lung type II pulmonary cells. During the 26th week of gestation, these cells commence the storage of glycogen, leading to the subsequent synthesis of pulmonary surfactant. This surfactant is vital for maintaining proper lung function.

Inulin

Inulin is the starch found in the plant of the family Compositae, like *Artichock, Dahlia, Dandelion,* etc. The polymerization of fructose molecules forms it. Inulin is the shortest polysaccharide with less than 100 β-D-fructose residues joined by β (1→2) glycosidic bond. It is used in physiological investigation for the determination of the rate of glomerular filtration.

Dextrins

Dextrins are the substances that are formed in the course of the hydrolytic breakdown of starch by the enzyme salivary amylase. On hydrolysis, they produce maltose, which is further hydrolyzed to glucose by the enzyme maltase. The partially digested starches are amorphous.

Dextrans

These are stored polysaccharides of yeast and bacteria. Their molecules are in the form of branched chains. In dextrans, D-glucose molecules are linked by α (1→6) glycosidic bond.

Cellulose

Cellulose is the most abundant carbohydrate present on earth. It forms the cell wall of plant cells and provides mechanical strength to the cells. It is formed of strong but elastic fibers. Each cellulose fiber comprises 10 to 15 thousand β D-glucose molecules joined by β (1→4) linkage. Some cellulose molecules together form the microfibril. They have excellent tensile strength. Cellulose is an essential component of the human diet. It adds bulk to intestinal contents and stimulates peristalsis in the inner wall. Cellulose is neither water-soluble nor easily digested by the amylase present in the digestive enzymes. Apart from this, cellulose nitrate is used in propellant explosives.

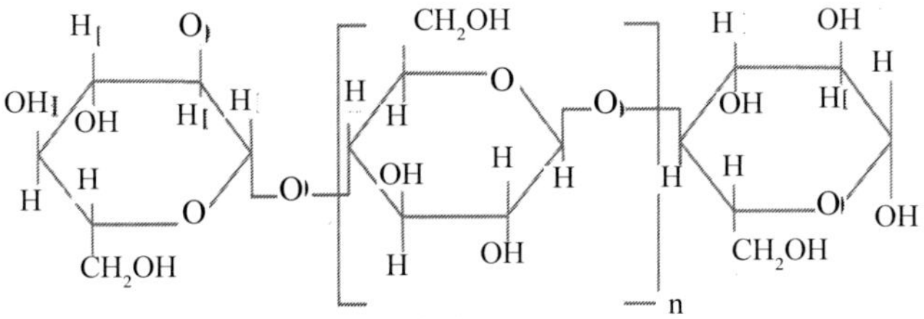

Cellulose

Chitin

Chitin ($C_{22}H_{54}N_4O_{21}$) is found in the exoskeleton of insects, other arthropods, and fungi's cell walls. Chitin, a modified structural carbohydrate containing nitrogen, is a N-acetyl-D-glucosamine polymer derivative of D-glucose joined through β (1→4) glycosidic bond. The discovery of chitin is attributed to the French scientist Henri Braconnot in 1811. Braconnot was the first to identify and describe this compound, which he extracted from mushrooms and found insoluble in sulfuric acid. He named it fungine. Another French scientist, Auguste Odier, also succeeded in isolating the compound from beetle cuticles and named it chitine, derived from the Greek word "chiton," meaning covering. Eventually, the compound became known as chitin. In 1929, the Swiss scientist Albert Hofmann, famous for his discovery of lysergic acid diethylamide (LSD), elucidated the structure of chitin.

Chitin is soft and skin-like but becomes hard due to some proteins and calcium carbonate deposition. It is insoluble in water and gives definite form to animals. Chitin shares structural similarities with cellulose. Since its discovery, chitin has found numerous applications in medicine and industry. Its non-toxicity, biodegradability (making it more environmentally friendly), and antimicrobial properties make it highly versatile. Its versatile nature has led to its utilization in various areas, such as the development of surgical stitches that dissolve over time and its inclusion as an additive in specific creams and cosmetics.

Peptidoglycans

The bacterial cell wall macromolecule called peptidoglycan is a heteropolysaccharide in which linear polysaccharide (glycan) chains are linked with tetrapeptide fragments (chains of amino acids). The bacterial peptidoglycan is a rigid structural polysaccharide that provides crucial structural support and safeguards against the risk of bacterial rupture during unfavorable environmental conditions. Peptidoglycan contains two components, i.e., polysaccharide and tetrapeptide, in which polysaccharide consists of repeating units of N-acetylglucosamine (NAG) and N-acetylmuramic acid (NAM) joined by β (1→4) linkage. A peptide bond links polysaccharides with tetrapeptides. Various peptide chains are cross-linked with tetrapeptides attached to NAM residues of polysaccharides. Gram-positive bacteria like Staphylococcus aureus have a pentaglycine chain in the cross-link, whereas gram-negative bacteria like E. coli lack the pentaglycine bridge.

Glycosaminoglycans (GAGs)

Glycosaminoglycans, also known as mucopolysaccharides, are heteropolysaccharides made up of disaccharide units, i.e., a hexose carbon sugar ring or a hexuronic acid bonded with a hexosamine. Additionally, one or both of the monosaccharide units in these chains contain sulfate or carboxylate groups. They are essential components of the extracellular matrix (blood vessels, cartilage, skin, bones, etc). The main mucopolysaccharides found in animals are hyaluronic acid, chondroitin sulfate, heparin, and keratin sulfate.

Hyaluronic acid

It is present in the extracellular matrix of all tissue, in the synovial fluid of bone joints, the cerebrospinal fluid of the brain, and the spinal cord. It consists of alternate units of D-glucuronic acid and N-acetyl D-glucosamine linked through a β (1→3) glycosidic bond to form a thread-like structure.

Chondroitin sulfate

It is found in the matrix of connective tissue, cartilage, tendon, and skin. It provides strength and elasticity to these structures. It comprises D-glucuronic acid and N-acetyl D-galactosamine 4-sulfate linked through a β (1→3) bond.

Heparin

Heparin is a polysaccharide and is a powerful antithrombic agent produced and stored by the mast cells. It consists of a chain of D-glucuronic acid 2-sulfate and N-sulfoglucosamine 6-sulfate linked through a β (1→3) linkage. Purified heparin is added to blood samples to avoid coagulation.

Dermatan sulfate

Dermatan sulfate is responsible for the flexibility of blood vessels, heart valves, and the skin. It contains L-iduronic acid and N-acetyl D-galactosamine 4-sulfate.

Keratin sulfate

Keratin sulfate is present in the connective tissue, cartilage, bone, cornea, and various horny structures formed of dead cells, horns, hair, hoofs, nails, and claws. It is formed from D-galactose and N-acetyl D-glucosamine 6-sulfate.

Agar and Agarose

Agar and agarose are gelatinous, non-nitrogenous mucopolysaccharides obtained from the cell walls of red algae. Agar is a polymer of D-glucose and D-galactose with sulfate esterification. It is used as a solidifying agent in culture media. At the same time, agarose is the polymer of D-galactose and 3, 6-anhydro-L-galactose and is used in electrophoresis.

Amino Sugars

Numerous chemical compounds known as amino sugars possess a sugar backbone with one of the hydroxyl groups substituted by an amine group. Over 60 amino sugars have been identified, and one of the most prevalent ones is N-Acetyl-D-glucosamine, a 2-amino-2-deoxysugar that serves as the primary constituent of chitin.

Hexosamines are a type of amino sugar that has a sugar derivative known as hexose. Some examples of hexosamines include glucosamine, galactosamine, fructosamine, and mannosamine. Glucosamine is naturally found in crustacean shells, animal bones, and fungi. It can also be produced synthetically and marketed as a dietary supplement in various forms, such as N-acetylglucosamine, glucosamine sulfate, or glucosamine hydrochloride, often combined with chondroitin, to help alleviate arthritis symptoms.

N-acetylglucosamine, derived from glucose, is an amino sugar with a molar mass of 221.21 g/mol and a chemical formula of $C_8H_{15}NO_6$. This amide compound plays a significant role in various biological structures, making it naturally abundant. It is a crucial component in the cell wall of bacteria, certain fungi, and the exoskeleton of specific insects known as chitin. Additionally, it can be found in the shells of crustaceans and forms part of the radula in mollusks, which aids in feeding. Moreover, N-acetylglucosamine is a significant constituent of cephalopods' beaks and contributes to the formation of glycosaminoglycans, glycolipids, and glycoproteins in biological membranes.

Galactosamine, on the other hand, is an amino sugar derived from galactose. It occurs naturally in certain glycoprotein hormones like luteinizing hormone and follicle-stimulating hormone.

Fructosamine is an amino sugar that is based on fructose. In humans and other animals, it forms when glucose combines with protein in blood serum, mainly albumin. This results in fructosamine, also known as glycated serum protein. The levels of glycated proteins in the blood can be monitored by measuring the amount of fructosamine in a blood sample, especially in individuals with diabetes.

Mannosamine, derived from mannose, is another type of amino sugar. In biological systems, D-mannosamine is a neuraminic acid component, a nonose or a 9-carbon monosaccharide. Neuraminic acid is one of the derivatives of sialic acids, which are widely found in animal tissues, particularly in glycoproteins and gangliosides. In humans, sialic acid is most abundant in the brain, which plays a role in neural transmission.

Glycolipids

A lipid molecule in the cell membrane can be covalently linked to a carbohydrate, typically an oligosaccharide. This type of carbohydrate is known as a glycolipid. Examples of glycolipids include glycosphingolipids and glyceroglycolipids. Additionally, glycolipids on the surface of erythrocytes determine human blood types. These biomolecular structures are located in the phospholipid bilayer of the cell membrane, with the carbohydrate part extending outward from the cell. The presence of glycolipids is essential for maintaining the stability of the plasma membrane, and they play a significant role in cell-to-cell interactions, such as cell adhesion to form tissues. Additionally, glycolipids are important for cellular recognition, which is crucial for immunological functions. A glycosphingolipid exemplifies a glycolipid, characterized by the linkage of a carbohydrate molecule and a sphingolipid molecule through a glycosidic bond. Upon hydrolysis, glycosphingolipids yield sugar, fatty acid, and sphingosine (or dihydrospingosine). Another instance of a glycolipid is a glyceroglycolipid, which comprises a glycerol backbone and at least one fatty acid. This category encompasses galactolipids and sulfolipids. The glycolipids present on the surface of erythrocytes determine four human blood types such as A, B, AB, and O. The oligosaccharide component of the glycolipid is responsible for determining the blood group antigen. For example, blood type A contains N-acetylgalactosamine, blood type B contains galactose and blood type AB possesses both sugars, while blood type O lacks both carbohydrate moieties.

Glycoproteins

A glycoprotein is a type of protein covalently linked to one or more carbohydrate residues through a process called glycosylation. Glycoproteins consist of various carbohydrate constituents, including β-D-glucose, β-D-galactose, β-D-mannose, α-L-fucose, N-acetylglucosamine, N-acetylgalactosamine, N-acetylneuraminic acid, and xylose. These carbohydrates are attached to the protein through O-glycosylation, where they bind to the OH group either of serine or threonine, or through N-glycosylation, where they bind to the amide NH_2 group of asparagine. Naturally occurring glycoproteins can be found in collagen, mucins, transferrin, ceruloplasmin, immunoglobulins, antibodies, histocompatibility antigens, hormones, enzymes, proteins involved in cell-to-cell interactions, lectins, selectins, protein receptors involved in hormone and drug action, calnexin, calreticulin, specific regulatory proteins for growth and development, and certain surface protein membranes of platelets involved in thrombosis.

Functions of Carbohydrates

Carbohydrates are widespread substances in plant and animal cells. Carbohydrates from skeletal structures serve as food reserves in plants, arthropods, and animals. They are significant sources of energy needed for a variety of metabolic functions; the energy that they provide is obtained by oxidation.

- Carbohydrates are used as readily available energy by living organisms to power cellular reactions. They are the most accessible and most prevalent dietary source of energy for every living organism (4 kcal/gram).
- Carbohydrates are the primary energy source in many animals and are also immediate means of energy. Glycolysis/Kerb's cycle breaks down glucose to produce ATP.
- They act as energy reserves, fuels, and metabolic intermediaries. It is stored in animals as glycogen and in plants as starch.
- Instead of proteins, stored carbohydrates serve as an energy source.
- They serve as structural and defensive components in plant and microbial cell walls.

- Carbohydrates play a crucial role as fundamental components in synthesizing fats and proteins.
- Carbohydrates play a crucial role in regulating nerve tissue and serve as a vital energy source for the brain.
- Carbohydrates form surface antigens, receptor molecules, vitamins, and antibiotics by interacting with lipids and proteins.
- Formation of the RNA and DNA structural frameworks (ribonucleic acid and deoxyribonucleic acid).
- Interconnected carbohydrates facilitate cell-cell communication and interactions between cells and their surrounding environment. These complex molecules are important mediators in relaying signals and facilitating various cellular processes.
- They are a vital component of connective tissues in mammals.
- Fiber-rich carbohydrates aid in the prevention of constipation.
- They also aid in the modulation of the immunological system.

Summary

- Carbohydrates are the polyhydroxy aldehydes or ketones or substances.
- The green plants synthesize carbohydrates through the process of photosynthesis.
- Carbohydrates are classified into three classes, i.e., monosaccharides, oligosaccharides, and polysaccharides.
- Monosaccharides are the simplest carbohydrates that cannot be hydrolyzed further to form simple forms.
- Monosaccharides are the fundamental building blocks of carbohydrates and the most basic form of sugar.
- Oligosaccharides are complex sugars that, upon hydrolysis, produce two to ten molecules.
- Disaccharides are oligosaccharides that produce two molecules of monosaccharides upon hydrolysis.
- Through glycosidic linkage, polysaccharides are formed by joining more than ten monosaccharide units.
- Storage polysaccharides are reserved energy sources, as glucose is the main energy source for cell metabolism.
- Isomers are molecules with the same chemical formula but different structural properties.

- D-glucose and D-galactose are epimers that differ at C4 carbon only.
- Glyceraldehyde is a significant molecule and standard of reference for the field of biochemistry.
- Glucose, also known as dextrose or blood sugar, is a widely recognized monosaccharide.
- D-glucose is the naturally occurring type of glucose.
- Fructose, also known as fruit sugar or levulose, is a naturally occurring ketohexose.
- Galactose, also known as milk sugar, is an aldohexose monosaccharide sugar.
- Maltose, a disaccharide, is formed by the linkage of two -D-glucose molecules through a glycosidic bond.
- Isomaltose is a disaccharide with the molecular formula C12H22O11 that comprises two glucose units.
- Lactose, commonly known as milk sugar, is a disaccharide that can be found in the milk of various mammals.
- Cellobiose is a reducing sugar formed by the combination of two β-glucose molecules through a β (1→4) bond.
- Starch is a mixture of two compounds, amylose and amylopectin.
- Glycogen is a branched polysaccharide composed entirely of α-D-glucose units.
- Cellulose is the most abundant carbohydrate composed of β-D-glucose molecules joined by β (1→4) linkage.

Long Type Questions

1. What are carbohydrates, and illustrate their classification.
2. Write a short note on important trioses and tetroses.
3. Elaborate glucose and fructose and describe their properties, uses, and different structures.
4. Write a short note on galactose, mannose, maltose, isomaltose, lactose and sucrose.
5. Write a short note on starch, glycogen, inulin, dextrins, cellulose, and chitin.
6. What are amino sugars?
7. Mention the functions of the carbohydrates.

Fill in the Blanks

1. Carbohydrates are the polyhydroxy ____________or _____________
2. The general formula of carbohydrates is ______________
3. Carbohydrates can be divided into three distinct groups, ____________, _______________, and __________________
4. Monosaccharides are _______ that ______________be hydrolyzed further to form simpler.
5. Oligosaccharides are ______________ sugars that, upon hydrolysis, produce __________ molecules of either distinct or identical monosaccharides.
6. Polysaccharides have more than __________ monosaccharide units joined by ____________
7. _____________ and _____________ are storage polysaccharides.
8. _____________ and _____________ are the important structural polysaccharides.

3

Lipids

Lipids are a class of organic compounds composed of hydrogen (H), carbon (C), and oxygen (O) atoms. These classes of biomolecules come together to create the fundamental building blocks that shape and govern the structure and functionality of living cells. The term lipid was used by the German biochemist Wilhem Bloor in 1943. Lipids are heterogeneous groups of substances that are insoluble in polar (water) and soluble in non-polar (ether, benzene, and chloroform) solvents. Lipids are the esters of fatty acids and their derivatives. Lipids play a crucial role in both the plant and animal domains. They are widely distributed in plants, but lipids are primarily stored in animal adipose tissues.

Classification of lipids

Lipids are classified into four different groups, as shown below:

1. Simple lipids
2. Complex lipids
3. Derived lipids and
4. Miscellaneous lipids

1. Simple lipids

Simple lipids are esters of fatty acids with alcohols. Simple lipids include

a) **Fats and oils** These are the esters of fatty acids with glycerol as an aliphatic alcohol and are neutral like triglycerides. They differ in their physical appearance, as they can be present in both solid (fats) as well as liquid (oils) state at room temperature. Depending upon the number of fatty acids in the ester, the glycerides may be monoglycerides, diglycerides, or triglycerides.

Triglycerides

Triglycerides are also known as neutral fats. They are fatty acid esters formed by bonding three molecules of the same or distinct nature of fatty acids with the alcohol (glycerol). For instance, tristearin comprises three molecules of stearic acid, whereas oleo-stearin consists of one molecule of oleic acid and two molecules of stearic acid. Therefore, the terminology of these compounds is based on their fatty acid constituents.

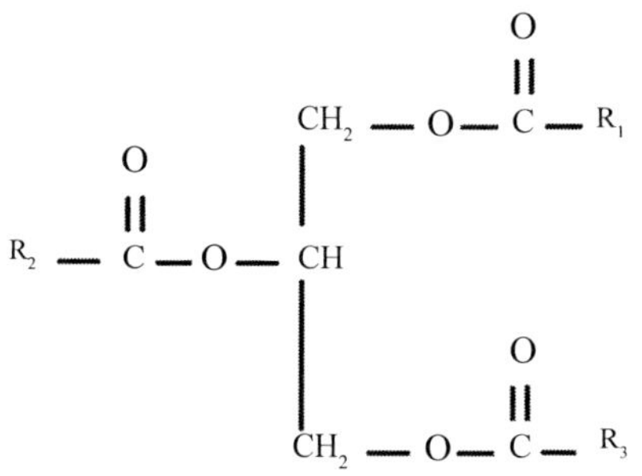

$$
\begin{array}{l}
\qquad\qquad\qquad\qquad\quad\;\; O \\
\qquad\qquad\qquad\qquad\quad\;\; \| \\
\qquad\qquad\qquad CH_2 - O - C - R_1 \\
\qquad\;\; O \qquad\quad | \\
\qquad\;\; \| \qquad\quad\; | \\
R_2 - C - O - CH \\
\qquad\qquad\qquad\quad | \qquad\qquad O \\
\qquad\qquad\qquad\quad | \qquad\qquad \| \\
\qquad\qquad\qquad CH_2 - O - C - R_3
\end{array}
$$

Structure of triglyceride

The structure of triglycerides can differ significantly depending on the fatty acids involved. These fatty acid chains can have varying lengths, with 16, 18, and 20 carbons being the most commonly observed. When the alcoholic group of glycerol reacts with the carboxylic group of long-chain fatty acids, a multitude of triglycerides can be formed. For example, if one ester bond is created, it forms a monoglyceride or monoacylglycerol. Similarly, when two carboxylic acid molecules react with glycerol, it generates two ester bonds, forming a diglyceride or diacylglycerol. Lastly, if three carboxylic acids react with a glycerol molecule, it creates three ester bonds, thus forming a triglyceride or triacylglycerol.

Hydrolysis of triglycerides

Under specific conditions such as acid and heat or in biological settings, triglycerides (fats) can undergo hydrolysis, producing glycerol and three fatty acids. Another widely utilized approach for the hydrolysis of triglycerides involves using a base, which is commonly employed on a large scale. This method, referred to as saponification, encompasses a single-step reaction that concurrently involves hydrolysis and neutralization, ultimately leading to soap formation.

Role of triglycerides in the body

To fuel cells that require energy, the body utilizes triglycerides. These naturally occurring compounds are present in the blood and stored as fat deposits, mostly in adipose tissue. Individuals who are overweight, lack physical activity and involve excessive consumption of carbohydrates or alcohol are at a higher risk of developing elevated triglyceride levels. However, an excess of triglycerides can give rise to difficulties in the body, potentially leading to serious health issues. Increased levels of triglycerides are closely associated with metabolic syndrome, a group of diseases that heighten the susceptibility to diabetes, obesity, stroke, and cardiovascular ailments.

b) **Waxes** Waxes are esters of fatty acids and have high molecular weight monohydroxy alcohol instead of glycerol. Chemically, waxes are inert because there are no double bonds in their hydrocarbon chains and are highly insoluble in water. Waxes are of great importance to living beings as they serve numerous functions, including water repellant, candle making (bee waxes), polishes, adhesives, and cosmetics.

2. Complex lipids

These are the esters of fatty acids and alcohol with additional compounds, such as phosphoric acid, sugar, proteins, etc. They are classified as phospholipids, glycolipids, and lipoproteins.

Phospholipids

Phospholipids are complex lipids encompass phosphoric acids, nitrogenous bases, alcohol and fatty acids. Phospholipids exhibit a hydrophilic head and a hydrophobic tail. These particular lipids hold significant importance as they serve as major cell membrane building blocks and contribute to the membrane's fluidity as they collectively form the bilayer. There exist two distinct categories of phospholipids, namely glycerophospholipids and sphingophospholipids.

Glycerophospholipids

Glycerophospholipids, being the predominant type, are commonly present in biological membranes. These phospholipids are composed of glycerol-based alcoholic molecules. A brief description of some glycerophospholipids is given as under:

Phosphatidic acid

It is the simplest and smallest phospholipid and an important intermediate in the biosynthesis of glycerophospholipids. The major phosphoglycerides are the derivatives of phosphatidic acid. Diacylglycerol kinase catalyzes the formation of phosphatidic acid from diacylglycerol.

Phosphatidylcholine

Also known as lecithin, it comprises one glycerol molecule, two fatty acids, and one phosphoric acid to which nitrogenous base choline is attached. It is the most abundant group of phospholipids present within the plasma membrane. It is widely distributed in animal cells, having both structural and metabolic functions as it is for utilization and transport of other lipids.

Phosphatidylethanolamine

It is also known as cephalin and is similar to lecithin, except it possesses ethanolamine instead of choline.

Phosphatidylserine

Another phospholipid that contains serine amino acid instead of ethanolamine. It is present in animal tissue and plays a vital role in cell signaling, particularly apoptosis.

Phosphatidylinositol

This type of glycerophospholipid is attached to an inositol molecule. It is an essential component of the cell membrane as it plays a crucial role in cell signaling.

Plasmalogens

Another glycerophospholipid resembles lecithins and cephalins but bears an ether linkage between a carboxylic group of fatty acids and C-1 of glycerol instead of the ester linkage in most phospholipids. It is found in myelin, red blood cells, skeletal muscle, and spermatozoa cell membranes.

Sphingophospholipids

These minor phospholipids contain an amino alcohol component (sphingosine) instead of glycerol and form amide linkage with fatty acids instead of ester linkage, thus forming ceramide. It includes the sphingomyelins that contain choline as a base instead of glycerol. Sphingophospholipids play a crucial role as they are constituents of myelin and are highly prevalent in the brain and nervous tissues.

Glycolipids

Glycolipids encompass a diverse group of intricate lipids comprising carbohydrates, fatty acids, sphingolipids, or a glycerol group. The term "glycolipids" is predominantly employed to characterize any compound in which carbohydrate residues are connected with the lipids through a glycosidic bond. These molecules are widely distributed in tissues, the brain, adrenals, kidneys, spleen, liver, leucocytes, thymus, lung, retina, egg-yolk, and fish sperm. Glycolipids serve as structural lipids primarily located on the outer surface of cell membranes. Additionally, it aids in the identification of an individual's blood group, contributing to the understanding of their genetic makeup. Moreover, it acts as a receptor on the surface of blood cells and facilitates various cellular processes, including immune responses. Glycolipids have different types, as mentioned below:

Cerebrosides

Cerebrosides are the simplest type of glycolipids or glycosphingolipids, predominantly present in the cell membranes of the neurons and the muscle cells. They consist of ceramide (sphingosine and fatty acids) and a monosaccharide (glucose or galactose) linked by a β-glycosidic bond. Their primary function is to act as an insulator and provide a protective coating to the nerve cells.

Gangliosides

A ganglioside is a compound consisting of a glycosphingolipid, which is made up of ceramide and oligosaccharide, along with one or more sialic acids (such as N-acetylneuraminic acid) attached to the sugar chain. Gangliosides are present in the membranes of all animal cells and are especially abundant in the plasma membranes of neurons. They have significant functions in regulating membrane proteins and ion channels, as well as in cell signaling and intercellular communication.

Lipoproteins

Lipoproteins are complex molecules formed through the combination of lipids and proteins and serve as crucial components in the body. Lipoproteins exist as spherical particles. These molecules act as carriers, facilitating the transportation of fats through the bloodstream throughout the body. They also help in the absorption and transport of dietary lipids within the small intestine. Moreover, lipoproteins are responsible for the movement of lipids from the liver to peripheral tissues and vice versa, a process known as reverse cholesterol transport. The lipoproteins can be categorized into five major classes, each possessing its distinct protein and lipid composition. These classes include chylomicrons, intermediate-density lipoproteins (IDL), very low-density lipoproteins (VLDL), low-density lipoproteins (LDL), and high-density lipoproteins (HDL).

3. Derived lipids

These are the substances produced by the hydrolysis of simple and complex lipids. It includes fat-soluble vitamins, steroid hormones, glycerol, eicosanoids, and ketone bodies.

4. Miscellaneous lipids

These substances possess lipid-like characteristics. This group includes terpenes, squalene, and carotenoids.

Sterols

Sterols are a group of lipid molecules with hydrophobic rings and exhibit a hydroxyl group positioned at the 3-position of the A-ring. Sterols are widely distributed in biological membranes. They can make up more than 50% of the lipid content in cells and can modify the structure and fluidity of these membranes. The chemical formula for sterol is $C_{17}H_{28}O$. Sterols have a heterocyclic ring structure consisting of six-membered three rings and five-membered one ring. Various forms can be found in different organisms, such as cholesterol, ergosterol, and stigmasterol. Cholesterol is the most abundant sterol in animal membranes. Ergosterol is another commonly found sterol in fungi and protists. Plant-derived sterols are called phytosterols and include compounds like campesterol, stigmasterol, and sitosterol.

Cholesterol

Cholesterol ($C_{27}H_{46}O$) is odorless, colorless, a 27-carbon compound, and a principal sterol. It is mainly found in the cells of animals and absent in plant cells; hence, it is known as animal sterol. It is abundantly present in the plasma membrane of animal cells and is a major constituent of lipoproteins. In mammals, cholesterol functions as a parent sterol, giving rise to vitamin D, bile acids, and steroid hormones (sex hormones and adrenocorticoids). Cholesterol plays a crucial role in animal cellular membranes by influencing their fluidity.

Adrenocorticoids

Adrenocorticoids are hormones synthesized and secreted by the adrenal cortex of the adrenal gland. Adrenocorticoids can be classified into three categories: glucocorticoids (cortisol), mineralocorticoids (aldosterone), and sex hormones (androgens and estrogens). These adrenocorticosteroids play crucial roles in various physiological and pharmacological processes, such as regulating glucose metabolism, aiding the body in coping with stress, and modulating immune function.

Sex steroids are a significant category of steroid hormones that play a crucial role in various reproductive/physiological processes. The group of sex steroids includes androgens, estrogens, and progestogens. Androgens are commonly referred to as male sex hormones due to their masculinizing effects. This group includes androstenediol, androstenedione, dehydroepiandrosterone, dihydrotestosterone, and testosterone. On the other hand, estrogens are known as female sex hormones because of their feminizing effects. Estrogens encompass estradiol, estetrol, estriol, and estrone. Progestogens, which are distinct from both androgens and estrogens, are another category of sex steroids. Among humans, progesterone is the only naturally occurring progestogen. In summary, sex steroids are a vital group of steroid hormones that are produced naturally in the gonads and adrenal glands.

Eicosanoids

Eicosanoids are 20-carbon-containing polyunsaturated fatty acids signaling molecules. Arachidonic acid, a polyunsaturated fatty acid, is the primary source of eicosanoids. Eicosanoids are further classified into three categories:

prostaglandins, thromboxanes, and leukotrienes. Eicosanoids are short-lived, localized paracrine hormones that affect the nearby cells. They contribute to processes such as inflammation and swelling at injury sites, labor induction, blood clotting, and asthma attacks.

Prostaglandins

Prostaglandins are a class of lipid autacoids. They are ubiquitously present in most tissues and organs and are synthesized by nearly all nucleated cells. The actions of prostaglandins vary depending on the specific site of their production, serving as signaling molecules that regulate a wide array of biological processes. When tissue damage or infection occurs, prostaglandins are produced, leading to the characteristic symptoms of inflammation, pain, and fever, which are crucial for the body's healing response. Additionally, prostaglandins also play a role in various reproductive functions, such as conception, luteolysis, menstruation, and parturition.

Leukotrienes

Leukotrienes, which are a group of eicosanoid inflammatory mediators, are produced in leukocytes through the enzymatic oxidation of arachidonic acid (AA) by the enzyme arachidonate 5-lipoxygenase. Leukotrienes exert their effects by interacting with specific receptors on target cells, thereby modulating various physiological and pathological processes. Leukotrienes have been proven to induce airway smooth muscle contraction, enhance vascular permeability, stimulate mucus secretion, and recruit and activate additional inflammatory cells within the airways of individuals who have asthma.

Thromboxanes

Platelets produce thromboxane, a type of lipid belonging to the eicosanoid family. These substances also referred to as vasoconstrictors and hypertensive agents, exert their effects by constricting blood vessels and promoting the formation of blood clots.

Fatty acids

Fatty acids are carboxylic acids of long-chain hydrocarbons represented by the formula $CH_3(CH_2)_n COOH$, where n represents the number of carbon atoms. Their carbon chain is long and unbranched. The carbon atoms in their chain may vary from 4 to 30 but typically range from 16 to 18. The carbon atoms in their chain are mostly even. Fatty acids have a carboxylic acid (-COOH) on one end and a methyl group ($-CH_3$) on the other end. The numbering of carbon atoms in fatty acids is started from the carboxyl carbon.

Saturated fatty acids

In saturated fatty acids, the hydrocarbon fatty acid chains have single bonds. Therefore, they have high melting points and exist as solids or semi-solids. Butyric acid (C4), lauric acid (C12), myristic acid (C14), palmitic acid (C_{16}), and stearic acid (C_{18}) acid are some of the important saturated fatty acids.

Unsaturated fatty acids

Unsaturated fatty acids have one or more double bonds. Two kinds of unsaturated fatty acids can be distinguished based on the number of double bonds they contain.

Monosaturated fatty acids

Unsaturated fatty acids containing a single double bond are called monounsaturated fatty acids, such as palmitoleic acid (C16) and oleic acid (C18).

Polyunsaturated fatty acids

Unsaturated fatty acids containing more than one double bond are called polyunsaturated fatty acids, e.g., linoleic acid (C18), linolenic acid (C18), and arachidonic acid (C20).

The most commonly used system for designing the position of the double bond in an unsaturated fatty acid is the delta (Δ) numbering system. The carboxyl carbon is designated as 1, and the position of the double bond is the number of the carbon atom from the carboxyl side, e.g., palmitoleic acid has 16 carbons and a double bond between carbon 9 and 10. It is designated as 16:1: Δ^9, or 16:1:9.

Based on body requirements, fatty acids are classified into two broad categories: essential and non-essential fatty acids.

Essential fatty acids

Our body cannot produce these fatty acids on its own, making it necessary to obtain them through diet. The three essential fatty acids are linoleic acid (C18), linolenic acid (C18), and arachidonic acid (C20). These polyunsaturated fatty acids are not synthesized as our body lacks the enzyme that can introduce double bonds beyond carbon number 9 (C-9).

Non-essential fatty acids

Conversely, our body can synthesize other fatty acids known as non-essential fatty acids. Palmitic acid (C18), stearic acid (C20), and arachidic acid (C22) are some of the non-essential fatty acids.

Common name	Systemic name	Abbreviation	Structure
Saturated fatty acids			
Accetic acid	Ethanoic acid	2:0	Ch_3COOh
Propionic acid	n-Propanoic acid	3:0	CH_3CH_2COOH
Butyric acid	n-Butanoic acid	4:0	$CH_3(CH_2)_2COOH$
Valeric acid	n-Pentanoic acid	6:0	$CH_3(CH_2)_3COOH$
Caproic acid	n-Hexanoic acid	8:0	$CH_3(CH_2)_4COOH$
Caprylic acid	n-Octanoic acid	10:0	$CH_3\ (CH_2)_6COOH$
Capric acid	n-Decanoic acid	12:0	$CH_3(CH_2)_8COOH$
Lauric acid	n-Dodenanoic acid	14:0	$CH_3(CH_2)_{10}COOH$
Myristic acid	n-Tetradecanoic acid	16:0	$CH_3(CH_2)_{12}COOH$
Palmitic acid	n-Hexadecanoic acid	18:0	$CH_3(CH_2)_{14}COOH$
Stearic acid	n-Octadecanoic acid	20:0	$CH_3(CH_2)_{16}COOH$
Arachidic acid	n-Cicosanoic acid	22:0	$CH_3(CH_2)_{18}COOH$
Behenic acid	n-Docasanoic acid	24:0	$CH_3(CH_2)_{20}COOH$
Lignoceric acid	n-Tetracosanoic acid	26:0	$CH_3(CH_2)_{22}COOH$
Unsaturated fatty acids			
Palmitolic acid	Cis-9-Hexadecenoic acid	16:1:9	$CH_3(CH_2)_5CH$- $CH(CH_2)_7COOH$
Oleic acid	Cis-9-Octadecenoic acid	18:1:9	$CH_3(CH_2)_7CH$- $CH(CH_2)_7COOH$
Linoleic acid	Cis,cis-9-12-Octadecenoic acid	18:2:9:12	$CH_3(CH_2)_4CH{=}CHCH$- $2CH{=}(CH_2)_2COOH$
Linolenic acid	All cis-9-12,15-Octade-cenoic acid	18:3:9:12:15	CH_3CH_2CH-$CHCH_2CH$-CH- CH_2CH-$CHCH_2CH$-CH $(CH_2)7COOH$
Arachidonic acid	All cis-5,8,11,14-Eicosatetraenoic acid	20:4:5:8:11:14	$CH_3(CH_2)_4CH{=}CHCH$- $2CH{=}CHCH_2CH{=}CH$- CH_2CH-$CH(CH_2)_3COOH$

Physiochemical properties of fatty acids

The physicochemical properties of fatty acids are given below.

a) Fatty acids are insoluble in water and soluble in organic solvents such as benzene, chloroform, and alcohol.

b) The melting point is directly proportional to the carbon chain length, i.e., with the increase in length of the carbon chain, the melting point increases and vice-versa.

c) They are bad conductors of heat.

d) At room temperature, long-chain saturated fatty acids assume a solid state, whereas long-chain unsaturated fatty acids remain in a liquid state.

e) Unsaturated fatty acids show cis-trans isomerism due to the presence of double bonds.

f) Trans-fatty acids have more boiling points than cis-fatty acids.

g) With an increase in temperature, the solubility of fatty acids also increases.

h) The melting point of saturated fatty acids is more than that of unsaturated fatty acids.

Chemical properties

a) **Hydrolysis** Hydrolysis by alkalis such as NaOH or KOH generates sodium or potassium salts of fatty acids. The process is called saponification.

b) **Hydrogenation** The process of fat hydrogenation involves the combination of unsaturated fat with hydrogen to convert it partially or entirely into saturated fat. By subjecting unsaturated fatty acids to high temperatures, hydrogen gas, and a metal catalyst, hydrogenated fats are formed through the process of artificial saturation.

c) **Hydrogenolysis** Hydrogenolysis involves the splitting of fats in the presence of hydrogen atoms. In this process, hydrogen is passed through fat under pressure in the presence of a copper-chromium catalyst.

d) **Halogenation** Halogenation involves treating unsaturated fatty acids with halogens such as chlorine and iodine. During this process, the unsaturated fatty acid takes up iodine or other halogens at their double bond site, indicating its unsaturated state.

e) **Rancidity** This process involves the complete or partial autoxidation or hydrolysis of fats and oils when exposed to air, light, moisture, or bacterial activity. This leads to the formation of short-chain aldehydes, ketones, and free fatty acids. When these reactions take place in food, they can cause unpleasant odors and flavors to develop.

f) **Emulsification** Emulsification. is the breakdown of large-sized fat molecules into smaller fat molecules. Various emulsifying agents, such as bile juice secreted by the liver, are used.

Biological significance of lipids

1. Principal food reservoir

Fasts are stored in the body as reserve food material. Triglycerides stored in adipocytes are the primary fat reservoirs.

2. Rich source of energy

Body lipids are the potential predominant energy source, as 1 gram of fat provides 9.3 kilocalories of energy.

3. Lipids are structural constituents of cells

Lipids are the structural components of the membrane system of the cell membranes and regulate the permeability of membranes.

4. Lipids are the precursors of essential compounds

Lipids serve as the precursor of fat-soluble vitamins (A, D, E, K) and bile salts

5. Lipids help in hormone synthesis

Hormones like adrenocorticoids and sex hormones are synthesized from cholesterol.

6. Lipids as signaling molecules

Lipids possess the capability to directly bind or modify the activity or position of target proteins, thereby functioning as signaling molecules.

7. Protection, trauma, and heat loss

Subcutaneous fat deposits around the visceral organs and underneath the skin insulate the body against mechanical shock and excessive heat loss, especially in aquatic animals and those living in cold environments.

Analytical methods

1. Saponification number

The saponification number denotes the amount of KOH, measured in milligrams, necessary to saponify a single gram of fat or oil. It serves as a measure of the molecular size of fatty acids.

2. Acid number

The acid number denotes the amount of KOH, measured in milligrams, necessary to neutralize the free fatty acid present in one gram of fat

3. Polenske number

The Polenske number is a term used to describe the quantity of 0.1 N KOH in milliliters that is needed to neutralize the insoluble fatty acids derived from 5 grams of fat.

4. Reichert-Meissl number

The number of ml of 0.1 N KOH required to neutralize the volatile fatty acid (separated by saponification, acidification, and steam distillation) contained in 5g of fat. It is used to measure the purity of butter.

5. Iodine number

The iodine number refers to the quantity of iodine (in grams) that is absorbed by 100 grams of fat. It serves as a measure of the degree of unsaturation in the fat.

6. Acetyl value

The potassium hydroxide (KOH) amount in mgs is needed to neutralize the acetic acid produced from 1g of acetylated fat or oil saponification. This value is used to quantify the number of alcoholic groups present in the fat.

Summary

- Lipids, being organic molecules abundant in energy, are essential for powering a wide range of biological activities.
- In contrast to their insolubility in water, lipids display solubility in non-polar solvents.
- Fatty acids are organic acids that typically possess lengthy aliphatic tails in the form of long chains. These chains can be either saturated or unsaturated.
- Triglycerides, classified as lipids, comprise a glycerol molecule and three fatty acid molecules.
- The presence of high triglyceride levels in the body can play a role in the progression of arteriosclerosis, a condition marked by the narrowing and hardening of arteries.
- A non-saponifiable lipid is resistant to hydrolysis, meaning it cannot be broken down into smaller molecules. Examples of non-saponifiable lipids are cholesterol and prostaglandins.
- A saponifiable lipid is composed of ester groups, which allow it to undergo hydrolysis when exposed to a base, acid, or enzymes. This category includes various lipids such as waxes, triglycerides, sphingolipids, and phospholipids.
- Within the cell membrane, cholesterol holds great significance as an essential lipid. It is categorized as a sterol, resulting from the combination of steroids and alcohol. The liver plays a significant role in cholesterol synthesis within the human body.
- The liver utilizes bile to assist in the breakdown of fats into fatty acids and glycerol, a process facilitated by the enzyme lipase.
- Prostaglandins and leukotrienes are examples of eicosanoids, which are lipids known for their biological activity. These molecules have been associated with both inflammation and cancer.

Long Type Questions

1. Describe lipids along with various types of lipids.
2. Describe the biological significance of lipids.
3. What are fatty acids? Describe different types of fatty acids.
4. Describe the physiochemical properties of fatty acids.
5. Define the following terms:
   a. Rancidity
   b. Saponification
   c. Polenske number
   d. Acid number
   e. Reichert-Meissl Number
   f. Iodine number
   g. Acetyl number.

Fill in the Blanks

1. Lipids were discovered by ________in the year________
2. Linoleic acid is _______fatty acid.
3. ______ are simple lipids containing one molecule of fatty acid and one molecule of high molecular weight monohydroxy alcohol.
4. Sterol and cholesterol are _______ lipids.
5. Prostaglandins, thromboxanes, leukotrienes, and lipoxins are the examples of __________

4

Amino Acids

Amino acids are organic molecules made up of carbon (C), hydrogen (H), oxygen (O), and nitrogen (N). However, sulfur (S) and selenium (Se) are also found in some cases. Amino acids are essential organic compounds that come together to create proteins, thus making them the fundamental building blocks of proteins. These biomolecules play vital roles in various biological and chemical processes within the human body and are essential for the growth and development of an individual.

Amino acids consist of an acidic group (-COOH), a basic group ($-NH_2$), and a side chain (R) or a hydrogen (H) atom, all bonded to the carbon atom. The features of amino acids are due to the side chain because they differ only in their side chain. All amino acids have the same basic structure except proline, which has a cyclic structure.

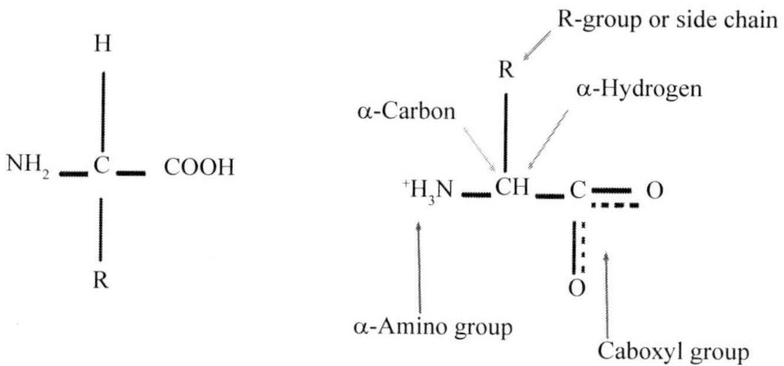

Various forms of amino acids like alpha (α), beta (β), gamma (γ) or delta (σ) form are found. The position of the amino group ($-NH_2$) determines the form of the amino acid, i.e., if the NH2 group is attached to the α-carbon, it is called α-amino acid; if β-carbon, it is called β-amino acid, and so on. Mostly, α-Amino acids are found as the numbering of amino acids is done from the carboxylic group (-COOH).

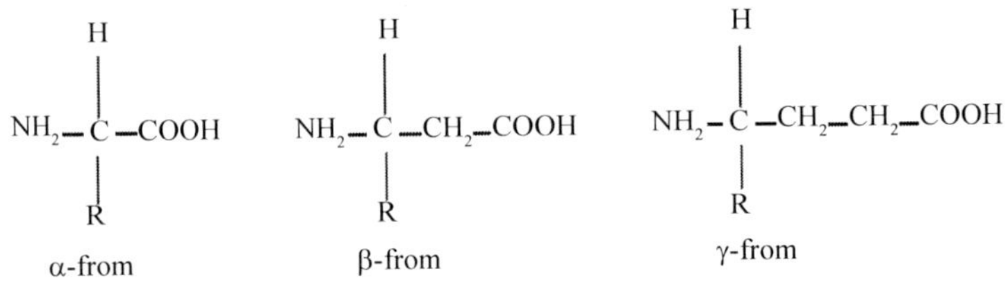

$$NH_2-\underset{R}{\overset{H}{C}}-COOH \qquad NH_2-\underset{R}{\overset{H}{C}}-CH_2-COOH \qquad NH_2-\underset{R}{\overset{H}{C}}-CH_2-CH_2-COOH$$

α-from β-from γ-from

In α-amino acids, the carboxylic, amino, and side chains are attached to the same carbon called α-carbon. More than 300 different types of amino acids are found. Based on this α-carbon they are differentiated into two categories:

Asymmetrical amino acids or chiral amino acids

These amino acids consist of asymmetrical or chiral α-carbon. The α-carbon atom is attached to four distinct atoms or groups of atoms. The main features of this group are:

1. Optically Active.
2. Have absolute D or L configuration.
3. Non-superimposable on their mirror images called enantiomers.
4. Have identical physical properties and chemical properties.
5. An example is all amino acids except glycine.

Symmetrical amino acids or achiral amino acids

These amino acids consist of symmetrical or achiral α-carbon. The α-carbon atom in this particular group is not attached to four distinct types or groups of atoms. The main features include

1. Optically inactive.
2. No absolute D or L configuration.
3. Not non-superimposable on their images.
4. They have different physical and chemical properties.
5. An example is glycine amino acid.

Classification of amino acids

The distribution of amino acids into proteins or polypeptides during the translation process serves as the basis for classifying amino acids into standard or non-standard categories.

Standard amino acids

The 22 common α-amino acids, known as the standard amino acids, are present in almost all proteins. These are ribosomically incorporated into the polypeptide and are involved in protein synthesis. These amino acids exhibit variations in the structure of their side chains, which are attached to their α-carbon atoms. It is important to note that all the standard amino acids are classified as L-amino acids. These amino acids have codons like UUU and AAA that code for phenylalanine and lysine, respectively.

Non-standard amino acids

Non-standard amino acids denote amino acids that undergo chemical modifications after their incorporation into a protein, a process known as "posttranslational modification." Furthermore, they are enzymatically integrated into the polypeptide chain. Additionally, nonstandard amino acids include those that occur naturally in living organisms but are not produced during the translation process. It is noteworthy that the genetic code for these non-standard amino acids is absent. These amino acids possess a D-form. Examples are 4-hydroxyproline, 5-hydroxylysine, peptidoglycan, desmosine, ornithine, citrulline, homoserine, and creatinine.

Standard amino acids are further classified because of various features:

1. Chemical nature of side chain.
2. Nature of side chain, i.e., based on polarity.
3. Nutritional value.
4. Metabolic fates.

1. Classified based on the chemical nature of the side chain

The chemical nature of their side chains determines the categorization of the standard amino acids into eight groups.

i. Aliphatic amino acids: Amino acid containing aliphatic side chain.

S. No.	Name	Symbol		R Group
		3 letter words	1 letter word	
1	Glycine	Gly	G	-H
2	Alanine	Ala	A	$-CH_3$
3	Valine	Val	V	$-CH-(CH_3)_2$
4	Leucine	Leu	L	$-CH_2-CH-(CH_3)_2$
5	Isoleucine	Ile	I	$-CH(CH_3)-C_2H_5$

ii. Aromatic amino acids: These amino acids possess aromatic side chains.

S. No	Name	Symbol		R group
		3 letter words	1 letter word	
1	Tyrosine	Tyr	Y	$-CH_2-(C_6H_5-OH)$
2	Phenylalanine	Phe	F	$-CH_2-C6H_6$
3	Tryptophan	Trp	W	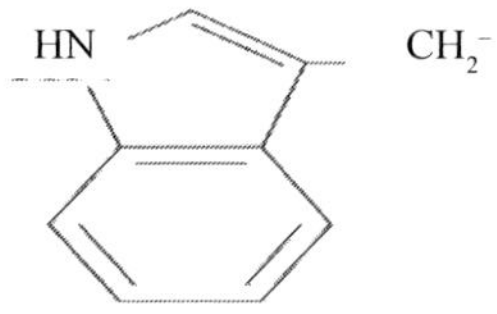 HN CH_2^-

iii. Hydroxyl amino acids: These amino acids bear a hydroxyl (-OH) group in their side chain.

S. No.	Name	Symbol		R group
		3 letter words	1 letter word	
1	Serine	Ser	S	$-CH_2-OH$
2	Threonine	Thr	T	$-CHOH-CH_3$
3	Tyrosine	Tyr	Y	---

iv. Sulfur amino acids: The sulfur (S) containing amino acids possess sulfur atoms in their side chain.

S. No.	Name	Symbol		R group
		3 letter words	1 letter word	
1	Cysteine	Cys	C	$-CH_2-SH$
2	Methionine	Met	M	$-CH_2-CH_2-S-CH_3$

v. Carboxylic amino acids: In these amino acids, a carboxylic (-COOH) group is in their side chain.

S. No	Name	Symbol		R group
		3 letter words	1 letter word	
1	Aspartate	Asp	D	$-CH_2-COOH$
2	Glutamate	Glu	E	$-CH_2-CH_2-COOH$

vi. Basic amino acids: This group contains an amino ($-NH_2$) group in its side chain.

S. No	Name	Symbol		R group
		3 letter words	1 letter word	
1	Lysine	Lys	K	$-CH_2-CH_2-CH_2-CH_2-NH_2$
2	Arginine	Arg	R	$-CH_2-CH_2-CH_2-NH-C(NH_2)=NH$
3	Histidine	His	H	CH_2 (imidazole ring: HN^+, NH)

vii. Amide amino acids: This group contains an amide ($-CONH_2$) group in their side chain.

S. No	Name	Symbol		R group
		3 letter words	1 letter word	
1	Asparagine	Asn	N	$-CH_2-CO-NH_2$
2	Glutamine	Gln	Q	$-CH_2-CH_2-CO-NH_2$

viii. Imino amino acids: This amino acid contains an imino (-NH) group in its side chain.

S. No	Name	Symbol		R group
		3 letter words	1 letter word	
1	Proline	Pro	P	Cyclic structure

2. Classified based on the nature of the side chain

Based on the polarity of the side chain, the amino acids are classified into two categories along with their sub-classes.

(i) Polar amino acids

Polar amino acids possess side chains that exhibit hydrophilic properties, making them soluble in water. These polar amino acids contain chemical groups in their side chains that are inherently polar, such as hydroxyl groups. Polar amino acids are further classified into charged and uncharged amino acids.

a) **Charged Amino acids** These amino acids carry a charge, either positive or negative. The R-groups of charged amino acids are electrically charged under physiological pH. They are further subdivided into two classes.

Positively charged amino acids

Conversely, amino acids with an amine functional group in their side chain are termed polar basic amino acids and exhibit an overall positive charge. This group includes amino acids like arginine (Arg), histidine (His), and lysine (Lys).

Negatively charged amino acids

Amino acids that contain carboxylic acid in their side chains are called polar acidic amino acids and display an overall negative charge. Glutamate (Glu) and aspartate (Asp) are examples of negatively charged amino acids, also known as polar acidic amino acids.

b) **Uncharged amino acids** These amino acids carry no overall charge. Serine (Ser), threonine (Thr), cysteine (Cys), asparagine (Asn), glutamine (Gln), and tyrosine (Tyr) are the specific amino acids that belong to this category.

(ii) Non-polar amino acids

Amino acids possess either long carbon chains or carbon rings as side chains, resulting in their bulky structure. These side chains are hydrophobic, meaning they repel water. Nonpolar amino acids are classified based on their chemical properties and do not carry any charge. The absence of polarity is due to the presence of nonpolar side chains like methyl or ethyl groups. Alanine (Ala), valine (Val), glycine (Gly), leucine (Leu), isoleucine (Ile), methionine (Met), proline (Pro), tryptophan (Trp), and phenylalanine (Phe) are examples of non-polar amino acids.

3. Classified based on nutritional value

The synthesis of proteins involves the utilization of α-amino acids. However, humans and other animals are not capable of synthesizing all the 20 amino acids, while plants and microorganisms can synthesize all the amino acids. Therefore, humans and other animals must obtain some preformed amino acids through their diet. As a result, such amino acids are divided into various classes, as mentioned below:

Non-essential amino acids

The non-essential amino acids are a set of amino acids that our body can synthesize. This group includes glycine (Gly), alanine (Ala), serine (Ser), cysteine (Cys), aspartic acid (Asp), asparagine (Asn), glutamic acid (Glu), glutamine (Gln), tyrosine (Tyr), proline (Pro), and arginine (Arg).

Essential amino acids

Essential amino acids are a group of amino acids that the body cannot produce and, therefore, must be obtained through dietary sources. These include valine (Val), isoleucine (Ile), leucine (Leu), lysine (Lys), methionine (Met), phenylalanine (Phe), tryptophan (Trp), threonine (Thr), and histidine (His). These amino acids play crucial roles in various physiological processes and are necessary for the proper functioning of the body.

Conditional essential amino acids

Usually, non-essential amino acids are synthesized in limited amounts, but during illness, our body cannot synthesize them adequately and, therefore, must be supplied through diet. Hence, they become conditionally essential. Arginine (Arg), cysteine (Cys), glycine (Gly), glutamine (Gln), tyrosine (Tyr), and proline (Pro) are some of the conditional essential amino acids.

Semi-essential amino acids

Some amino acids are synthesized by adults and not by children (infants) and hence are considered semi-essential amino acids. Arginine (Arg) and histidine (His) are semi-essential amino acids.

4. Classified based on metabolic fate

Based on metabolic fate, amino acids are classified into three categories.

Glucogenic amino acids

Within the process of amino acid degradation, specific amino acids possess the capability to act as a precursor for glucose synthesis by providing a carbon skeleton, hence being referred to as glucogenic amino acids. These include glycine (Gly), valine (Val), alanine (Ala), serine (Ser), cysteine (Cys), methionine (Met), aspartic acid (Asp), asparagine (Asn), glutamic acid (Glu), glutamine (Gln), arginine (Arg), histidine (His), and proline (Pro).

Ketogenic amino acids

The amino acids that can be degraded to acetyl-CoA and serve as precursors for fatty acid synthesis are known as ketogenic amino acids, including lysine (Lys) and leucine (Leu).

Amphibolic amino acids

The amino acids that are both glucogenic and ketogenic are known as amphibolic amino acids. Phenylalanine (Phe), tyrosine (Tyr), tryptophan (Trp), threonine (Thr), and isoleucine (Ile) are both Glucogenic and ketogenic.

Properties of amino acids

The physical properties of amino acids are shown below:

1. Optical property

All amino acids are optically active except glycine amino acid due to the presence of a chiral carbon atom. Optically active means that they can rotate the plane of plane polarised light (monochromatic light) either right (called Dextrorotatory or simply d) or left (called laevorotatory or simply l). It is a quantitative measurement, and the instrument used is a polarimeter. The measurement of the optical activity is quantified by determining the angle of rotation. This quantitative measurement of optical activity is called Specific rotation (α), and it depends upon the path length of light(l), the concentration of compound (C), wavelength of plane polarised light (D = 589nm), and temperature (T = 25°C).

2. Melting point

Amino acids have a higher melting point (above 200°C).

3. Solubility

Amino acids are generally soluble in water and are slightly soluble in alcohol but are insoluble in other organic solvents.

4. Color and taste

Amino acids are colorless, crystalline solids with or without taste.

5. Absolute configuration

Two absolute configurations exist, i.e., the D-L system (mostly) and the R-S system. In the D-L system, an amino acid with chiral carbon exists in two stereoisomers, known as enantiomers (D/L). In this system, we use the hydroxyl (-OH) group as a reference group of the reference molecule glyceraldehyde. If the –OH group is present on the right side, it is called the D-form; if the left side, it is called the L-form. In the case of amino acids, we consider the NH_2 group as a reference group to determine the absolute configuration. When the NH_2 group is on the right side, it is known as the D-form, whereas the presence of the NH_2 group on the left side indicates the L-form.

In the case of the R-S system (R-Rectus=clockwise rotation and S-Sinister=anticlockwise rotation), we determine the R/S isomers based on atomic number while neglecting the atoms or molecules with low atomic number. According to the D-L system, all amino acids used for protein synthesis have an L-form, and according to the R-S system, all amino acids have an S-form (2S), except cysteine, which has an R-form.

6. Amphoteric nature

As we know, amino acids contain both acidic (-COOH) and basic (-NH2) groups and are mostly dissolved in water. When an amino acid is immersed in water, it possesses the capacity to act as an acid by donating a proton or as a base by accepting a proton. Thus, an amino acid in solution exists as a Zwitterion or dipolar ion, i.e., an ion that acts as either acid or base. Hence, amino acids are regarded as amphoteric molecules or ampholytes. A zwitterion is a hybrid molecule containing both positive and negative ionic groups; hence, it is also called a dipolar ion.

At acidic/low pH (high H^+ concentration), the carboxyl group accepts a proton and becomes uncharged; thus, the overall charge of amino acid becomes positive (cation). Similarly, at alkaline/high pH (low H^+ concentration), the amino group donates a proton and becomes uncharged; thus, the overall charge of amino acid becomes negative (anion). Hence, each amino acid has its characteristic pH. The pH at which the overall net charge of amino acid becomes zero is called Isoelectric pH (PI). At PI, the amino acids exist as a Zwitterion.

7. Absorption

Standard amino acids absorb electromagnetic radiation (UV-radiation) due to the presence of aromatic ring structure in three aromatic amino acids: phenylalanine (Phe), tryptophan (Trp), and tyrosine (Tyr). The maximum absorption takes place at 280nm. Due to these three amino acids, the proteins absorb at 280 nm.

Chemical properties

The chemical properties of amino acids are due to the two groups, i.e., acidic (-COOH) and basic (-NH2) groups.

1. Chemical reaction due to amino group

i. The amino acid can form salts due to the amino group.

ii. The amino group reacts with ninhydrin (strong oxidizing agent) and forms a blue, purple, or pink-colored complex called Ruhemann's purple/Diketohydrin. This reaction is used for both quantitative and qualitative analysis of amino acids.

iii. Amino groups are used in transamination reactions to form new amino acids and oxidative deamination to form ammonia.

iv. The amino group is also used for Urea formation.

2. Chemical properties due to carboxylic group

i. Like amino groups, carboxylic groups can form salts.

ii. The carboxylic group is used in decarboxylation reactions to form amines.

iii. The carboxylic group also reacts with ammonia to form amides.

Peptide bond

A peptide bond, also called an amide linkage/bond, is a covalent chemical bond that holds amino acids together to form a peptide or polypeptide. Peptide bond formation is a condensation reaction/nucleophilic acyl-substitution reaction/ dehydration reaction. During peptide bond formation, the unpaired electron pair of the α-amino-group of one amino acid reacts with the α-carboxyl-group of another amino acid, resulting in the combination of these two groups. Water molecule is removed during peptide bond formation. The linked amino acids are now called amino acid residues.

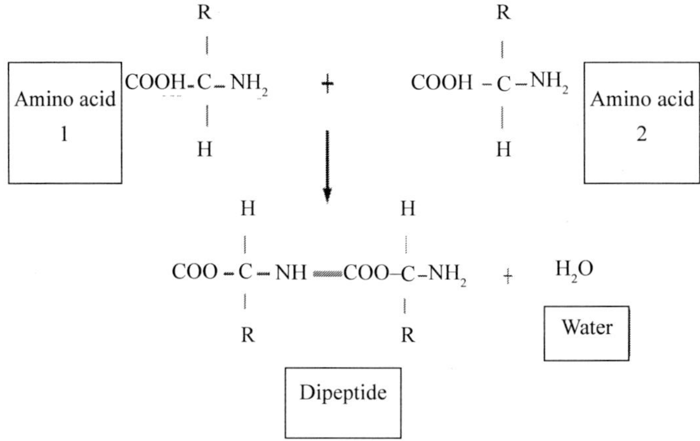

Characteristics of Peptide bond

1. It has a partial double bond character (about 40% double bond character) due to resonance.
2. The peptide bond is stronger than the normal C-N bond.
3. The peptide bond is shorter (1.33Å) than normal C-N bond.
4. The peptide bond has a planar nature.
5. A peptide bond is a rigid bond with no free rotation and only restricted rotation.
6. The peptide bond shows maximum absorption at 190 nm.

7. The peptide bond has two configurations based on the angle of rotation (ώ), i.e., Cis (ώ=0^0) and Trans (ώ=180^0). Mostly, the 'Trans' configuration has been found.
8. Peptide bond formation is an endergonic reaction during bond formation; energy is required).

Ramachandran plot

As we know, the peptide bond is rigid, and no free rotation is possible, but rotation is permitted about the N–C_α and the C_α–C bonds. Rotations about these bonds are called Dihedral torsion or conformational angles. The bond angle resulting from rotation about the N–C_α bond is denoted by *phi* (ϕ), and the bond angle resulting from rotation about the C_α–C bond is denoted by *psi* (ψ). The ϕ and ψ have different values, depending upon steric hindrance between the atoms and side chain of amino acids. The values are between -180^0 to +180^0. In 1963, G.N. Ramachandran determined the permit values of ϕ and ψ on the 2-D plot called the Ramachandran plot.

Deficiency of amino acids

The significance of amino acids as the building blocks of proteins cannot be overstated, as proteins are involved in nearly all essential life processes. Thus, including all the amino acids in the diet becomes crucial to ensure our bodies' proper functioning and maintenance. The absence or insufficiency of these amino acids can give rise to various disorders, including edema, diarrhea, depression, hypoglycemia anemia, insomnia, loss of appetite, fat accumulation in the liver, and problems related to the skin and hair. Furthermore, individuals experiencing amino acid deficiency may also exhibit symptoms such as headache, weakness, irritability, and fatigue.

Functions of amino acids

1. The presence of phenylalanine is essential for the proper functioning of the nervous system and has been linked to enhancing memory capabilities, thereby promoting a healthy cognitive state.
2. Valine plays a significant role in promoting muscle growth.
3. Threonine is crucial for enhancing the immune system's functions.
4. Tryptophan plays a fundamental role in the formation of vitamin B_3 and serotonin hormones. Serotonin, an essential hormone, governs our appetite, controls our sleep cycles, and boosts our overall mood.
5. Isoleucine is vital for creating hemoglobin, triggering the pancreas to produce insulin, and aiding in the transportation of oxygen from the lungs to various body tissues.

6. The utilization of methionine encompasses the treatment of kidney stones, the preservation of skin health, and the control of pathogenic bacteria invasion.
7. Leucine actively facilitates the process of protein synthesis and enhances the production of growth hormones.
8. Lysine is a critical player in the production of enzymes, antibodies, and hormones, which are crucial for overall health. Moreover, it also aids in the absorption and fixation of calcium in our bones, ensuring their proper growth and strength.
9. Histidine is vital for various enzyme processes and actively contributes to red and white blood cell formation.
10. The significance of glutamine in maintaining optimal brain function cannot be overstated, as it is indispensable for synthesizing DNA and RNA. These molecules form the basis of our genetic material and are crucial for various biological processes.
11. Glycine, on the other hand, plays a vital role in ensuring proper growth and function of cells and wound healing. It also acts as a neurotransmitter, facilitating communication between nerve cells.
12. Arginine, an essential amino acid, catalyzes the synthesis of hormones and proteins. It also facilitates the detoxification process in the kidneys and promotes wound healing.
13. In the production of thyroid hormones T3 and T4 and the synthesis of neurotransmitters and melanin, tyrosine assumes a pivotal function.

Identification of amino acids

The nature of amino acids is identified based on color. Various tests/reactions have been discovered to determine the nature of an amino acid.

Ninhydrin test

Ninhydrin reagent, also known as 2,2-dihydroxyindane-1,3-dione, is an oxidizing agent that finds application in identifying ammonia or primary and secondary amines. Upon reaction with unbound amines, Ninhydrin produces a complex of intense blue or purple color.

Xanthoproteic test

The Xanthoproteic test is a qualitative approach that can be employed to ascertain the presence of amino acids in a protein solution. This method involves the use of concentrated nitric acid as a reagent. A positive outcome of the test is indicated by the development of a dark yellow color, which arises from the neutralization of nitric acid with an alkali. The yellowish hue results from the presence of xanthoproteic acid, which is formed through the nitration of specific amino acids, with tyrosine and tryptophan being the most prevalent examples.

Sakaguchi test

The Sakaguchi test is specifically used to detect arginine in the protein solution. In this experiment, the Sakaguchi reagent, which is comprised of 1 naphthol and a small quantity of sodium hypobromite, is employed. Red-colored complex formation occurs due to the presence of guanidine group in arginine amino acid.

Hopkins-Cole test

The Hopkins-Cole test, also called the glyoxylic acid reaction, is used to detect the tryptophan amino acid in the protein solution. The Hopkins Cole reagent (consisting of glyoxylic acid) is used in this test. The addition of concentrated sulfuric acid forms two layers along with the purple ring between them.

Pauly's test

Pauly's reaction detects tyrosine or histidine in the protein solution. A red color is produced in this test by subjecting diazotized sulfanilic acid to a coupling reaction under alkaline conditions.

Millon's test

Millon's test detects tyrosine amino acid in the protein solution. In this experiment, Millon's reagent, which is composed of metallic mercury dissolved in nitric acid and subsequently diluted with water, is employed. Upon heating, an observable transformation takes place, resulting in the formation of a reddish-brown color.

Summary

- Amino acids are organic compounds that consist of fundamental amino groups (-NH_2) and carboxyl groups (-COOH). Hence, amino acids are regarded as the fundamental units from which proteins are constructed.
- The amphoteric nature of amino acids arises from their possession of both acidic and basic properties.
- Zwitterions refer to molecules that exhibit functional groups, wherein one group bears a positive electrical charge and another group bears a negative electrical charge. The overall charge of the zwitterion is maintained at zero. The minimum molecular weight is glycine, and the maximum is tryptophan.

- Optical activity is observed in all amino acids except for glycine.
- The maximum percentage of protein is found in leucine, and the minimum percentage in tryptophan.
- Methionine and cysteine are amino acids that contain sulfur.
- Highly non-polar (Hydrophobic) amino acid is isoleucine, and highly polar (Hydrophilic) is arginine.
- The most basic amino acid is arginine, and the most acidic is aspartate.
- The highest pI value is for arginine, and the lowest is for aspartate.
- The average molecular weight of amino acid is 128, and the average molecular weight of amino acid residue is 110.

Long type questions

1. Write an elaborate note on amino acids along with classification.
2. Write a short note on amino acids' physical and chemical properties.
3. What are the functions of amino acids?
4. Write short notes on the following: (i) Peptide bond, (ii) Ramachandran plot, (iii) Glucogenic and ketogenic amino acids.
5. What are the various tests used for amino acid identification?

Fill in the blanks

1. Amino acids are the building blocks of ____________
2. Amino acids consist of a _________ group, a _________ group, a _________, and a ____________ atom, all bonded to the carbon atom.
3. Based on the chemical nature of the side chain, the standard amino acids are classified into_________ groups.
4. Polar amino acids possess side chains that exhibit__________ properties, making them soluble in water.
5. Positively charged amino acids consist of _________, ____________, and ____________
6. _________ and ___________ are examples of negatively charged amino acids.
7. Amino acids synthesized in our body are called as _______________
8. Amino acid obtained through diet called ____________________
9. All amino acids are optically active except _____________ due to the presence of a chiral carbon atom.

5

Proteins

Proteins are composed of amino acids and are large molecules that play a crucial role in various biological processes. They are considered the fundamental building blocks of life due to their abundance in the body, constituting approximately 60% of the dry weight of cells. These complex biomolecules are intricately folded and twisted strands of amino acids, actively participating in vital life processes, including metabolism, locomotion, immune response intercellular signaling, and molecular identification. As macronutrients proteins are present in all living organisms and directly contribute to numerous metabolic pathways. It is important to note that proteins are unique to each organism, exhibit species specificity, and vary within different organs of the same organism. With 22 different amino acids, the properties of a protein molecule are determined by the specific arrangement of these amino acids. While plants can synthesize all the necessary amino acids for protein production, animals rely on external sources for certain amino acids. Proteins consist of amino acids connected through peptide bonds, formed by the linkage between the amino (NH_2) group of one amino acid and the carboxyl (COOH) group of another amino acid.

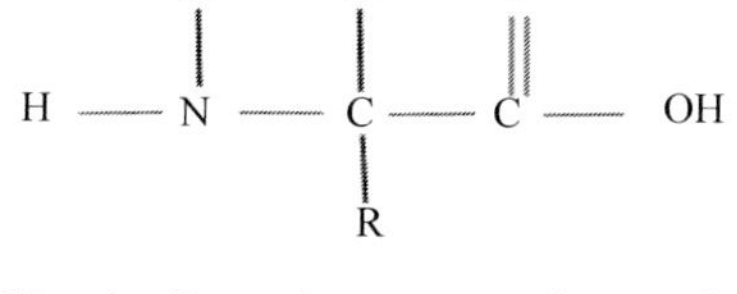

```
      H     H    O
      |     |    ||
H --- N --- C -- C --- OH
            |
            R
```

Fig. 1: General structure of an amino acid, where NH_2 = Amino group, COOH = carboxyl group, R = variable side chain.

Animal-based foods such as meat, poultry, fish, eggs, and dairy products are strongly advised as an abundant source of complete protein. However, plant-based foods such as fruits, vegetables, grains, nuts, and seeds may not offer all the essential amino acids the body requires.

Peptides

Peptides are short chains of amino acids that form proteins in conjunction with other peptides. Peptide chains are produced by linking two or more amino acids together through peptide bonds. Peptides are classified into various groups based on the number of amino acids they contain. For example, dipeptides are peptides composed of two amino acids, while tripeptides contain three amino acids, and polypeptides are those with more than ten amino acids. During protein synthesis, a special kind of amide bond is formed when the α-carboxyl group of one amino acid reacts with the α-amino group of another amino acid, peptide, or polypeptide. This process is known as translation and is classified as a dehydration synthesis reaction because it removes a water molecule.

Polypeptides

Polypeptides are peptides consisting of more than ten or twelve amino acids. Peptide bonds serve as the connections between amino acids during polypeptide formation. When multiple polypeptides are combined, they create a protein molecule with two terminals, i.e., N and C terminals. The N-terminal, also known as the amino-terminal, is the end of the polypeptide that contains a free amino group. On the other hand, the C-terminal, or carboxyl-terminal, is the end that possesses a free carboxyl group. The nucleotide sequences within the DNA molecule determine the sequence of codons in the mRNA, which in turn determines the specific sequence of amino acids in a polypeptide.

Classification of Proteins

Proteins are classified based on their chemical nature, structure, shape, and solubility. There are different types of proteins:

1. Simple proteins

These proteins are made up of only amino acid residue. When hydrolyzed, they break down into constituent amino acids. Simple proteins can be further divided into two categories:

a. **Fibrous proteins** Fibrous proteins possess elongated rope-like formations that exhibit strength and hydrophobicity, typically comprising predominantly a singular secondary structure. These structures offer support, shape, and external safeguarding to vertebrates and comprise fibrous proteins. Examples include keratin, elastin, and collagen.

Keratin

Keratin belongs to the scleroprotein family of fibrous structural proteins. The stratified squamous epithelium is composed of multiple layers of cells. Each layer of cells differs from one another. The cells in the deeper tissue are cuboidal in shape, which then transition to a polygonal shape and eventually flatten towards the outermost surface. These flattened cells on the surface contain keratin, a fibrous protein that renders them lifeless. Various structural components such as scales, horns, fur, feathers, nails, paws, calluses, hooves, and the outer layer of the skin are composed of this material.

Keratin serves several functions, including:

1. Protecting epithelial cells
2. Strengthening internal organs
3. Forming the protective outer layer of the skin
4. Strengthening and repairing hair
5. Maintaining the skin's elasticity and firmness
6. Controlling the growth and renewal of epithelial cells Moreover, keratin plays a vital role in maintaining the mechanical stability and integrity of epithelial cells and tissues.

Elastin

Elastin, an essential protein found abundantly in the human body, plays a crucial role in various tissues within the body that necessitate elasticity, such as the lungs, bladder, large blood vessels, and certain ligaments. Its primary function is to enable tissues to stretch and effortlessly revert to their initial size. Remarkably, elastin exhibits approximately 1000 times more flexibility than collagens.

Collagen

Collagen, the protein that is abundantly present in the body, acts as a fundamental framework for various parts, including the skin, bones, ligaments, and tendons. Besides its presence in the skin, bones, ligaments, and tendons, collagen can also be found in other areas such as teeth, cornea, and blood vessels. Its name is derived from the Greek word "kolla," which means glue, as it effectively serves as a binding agent connecting all structures. Approximately one-third of the body's protein composition comprises collagen, which contains specific amino acids like glycine, proline, hydroxyproline, and arginine. Although there are 28 different types of collagen in the human body, however, five types are commonly known and seen in supplements. Collagen plays several important roles, including facilitating the formation of fibroblasts in the dermis, which aids in the growth of new cells. It also contributes to replacing dead skin cells, provides a protective covering for organs, and imparts structure, strength, and elasticity to the skin. Additionally, collagen assists in the clotting of blood.

b. **Globular proteins** Globular proteins consist of amino acid sequences that are arranged irregularly. These proteins are soluble in water and can form colloids. Interestingly, globular proteins are often referred to as spheroproteins due to their spherical shape. Globular proteins commonly combine different secondary structures and possess a more rounded and hydrophilic morphology. Hence, a significant number of enzymes and regulatory proteins, including immunoglobulins, fall under the category of globular proteins. Notable examples of globular proteins include albumin, globulin, glutelin, and histones.

Albumin

Albumins are commonly found in blood plasma and belong to the family of globular proteins. Albumin is a primary and most abundant protein in the bloodstream, with a molecular weight of 66.5 kDa. It can be found in various bodily fluids such as blood, lymph, tissues, and egg whites. Albumin is soluble in water. It acts as a simple protein in both animal and plant physiological fluids and tissues. The most prevalent type of albumin is serum albumin. Serum albumin is present in blood serum, while α-lactalbumin is found in milk and ovalbumin in egg white. Ovalbumin constitutes approximately 50% of the proteins in egg white, and conalbumin represents 3-16%. However, albumin levels in seeds are extremely low, accounting for only 0.1-0.5% of their weight. The normal range for albumin levels typically falls between 3.4 to 5.4 g/dL. If the albumin level falls below this range, it could indicate malnutrition, liver disease, kidney disease, or an inflammatory disease. Conversely, elevated albumin levels may be attributed to acute infections, burns, or stress resulting from surgery or a heart attack. The function of albumin includes maintaining appropriate osmotic pressure, binding and transporting substances like hormones and drugs in the blood, and neutralizing free radicals.

Globulin

Globulins, a group of globular proteins, have molecular weights higher than albumins and are not soluble in pure water. However, they can dissolve in diluted salt solutions. The liver is responsible for producing some globulins, while the immune system generates others. Typically, normal human blood contains a concentration of globulin protein ranging from 2.6 to 4.6 g/dL. These proteins play vital roles in liver function, blood coagulation, and the body's defense against infections. Alpha, beta, and gamma globulins are the various types of globulins.

Glutelin

Glutelins, belonging to the prolamin protein family, are specifically found in the endosperm of certain grass family seeds. They play a vital role in the composition of gluten, a protein composite. These proteins can be dissolved in weak acids and alkalis and solidify when subjected to heat. Their molecular weights typically range from 40 to 50 kDa. Prominent examples include glutenin from wheat and oryzenin from rice. Glutenin is the most abundant among glutelins and is predominantly present in wheat. It significantly contributes to the dough's strength and elasticity. Glutelins are recognized as high-quality plant proteins that contain essential amino acids, making them easily digestible and absorbable compared to prolamins.

Histones

Chromosomes contain histones, which are the basic proteins that contribute to the organization and packaging of the DNA helix within the chromatin fiber in the nucleus. In eukaryotes, the DNA is highly condensed and tightly wound around histones, forming a chromatin thread that eventually condenses further to form chromosomes. By providing a structural framework, histones ensure that the chromosomes are compact enough to fit inside the nucleus. These proteins carry a positive charge and contain amino acids like arginine and lysine, which bind to the negatively charged DNA. The five primary types of histones are H_1, H_2A, H_2B, H_3, and H_4.

Function

1. Histones play a crucial role in the compaction and organization of DNA within chromatin fibers in the nucleus.
2. The presence of histones is essential for the efficient packaging of DNA into chromatin fibers. Moreover, they provide structural support to chromosomes.
3. By forming complexes with histone proteins, DNA can adopt a more condensed shape, resulting in the characteristic compact structure of chromosomes. Histones also play a significant role in controlling gene expression.

2. Conjugated proteins

These proteins can be found in combination with a non-protein moiety, such as sugars, lipids, phosphoric acid, metal ions, and nucleic acids. Some notable examples of such combinations include nucleoproteins, glycoproteins, phosphoproteins, lipoproteins, and metalloproteins.

a) **Nucleoproteins** Nucleoproteins are intricate combinations of nucleic acids and proteins. The nucleic acid component can either be DNA or RNA. A few well-known instances of nucleoproteins encompass chromosomes, ribosomes, and more. Chromosomes, specifically, are composed of DNA coiled around fundamental proteins known as histone proteins. Nucleoproteins are present in every living cell and serve multiple functions, such as providing structural support, regulating genes, and facilitating replication.

b) **Glycoproteins** Glycoproteins are proteins that are connected to carbohydrate groups through covalent bonds. The quantity of carbohydrate groups in different glycoproteins can vary greatly, with some containing less than 1% (like egg albumin) and others having as much as 80% (such as mucoproteins). When glycoproteins have a high amount of carbohydrates, they are called proteoglycans. These molecules are essential in the immune system as they help white blood cells move around the body, trigger immune responses, and recognize other cells. Additionally, they are involved in the production of mucus, which serves as a protective barrier for various organs in our body. Glycoproteins are vital for maintaining the overall health and functionality of our bodies. Naturally occurring examples of glycoproteins include collagen, mucins, transferrin, ceruloplasmin, mmunoglobulins, histocompatibility antigens, and various hormones.

c) **Phosphoproteins** Proteins that undergo modification by adding phosphate groups are known as phosphoproteins. Phosphoproteins play a crucial role in signal transduction pathways, which are essential for various cellular processes. Through phosphorylation, the activity of numerous enzymes is regulated, leading to significant changes in their functionality. Casein, a protein derived from milk, and ovo-vitellin of egg yolk are examples of this protein type.

d) **Lipoproteins** Lipoproteins are protein-lipid complexes in various body parts such as the brain, egg, milk, and plasma. These complexes consist of cholesteryl ester and triacylglycerol, surrounded by a single layer of amphipathic phospholipid and cholesterol molecules. Lipoproteins are crucial in carrying and delivering fatty acids, triacylglycerol, and cholesterol to different organs' target cells. There are five main types of lipoproteins, categorized based on their density and the ratio of proteins to lipids. These types include chylomicrons, very low-density lipoproteins (VLDL), low-density lipoproteins (LDL), intermediate-density lipoproteins (IDL), and high-density lipoproteins (HDL).

e) **Metalloproteins** Metalloproteins, also known as chromoproteins, are a diverse group of proteins that play crucial roles in cellular functions. These proteins contain a metal ion cofactor, which enables them to perform a wide range of tasks. They contribute to vital processes such as protein storage and transport, enzyme activity, signal transduction, and even the development of infectious diseases. Examples of metalloproteins include hemoglobin, hemocyanin, cytochromes, and flavoproteins.

Hemoglobin

Hemoglobin (Hb) is a spherical protein that can be found exclusively in red blood cells (RBCs) or erythrocytes. It is synthesized within the bone marrow cells and undergoes a transformation process to become a constituent of red blood cells. Hemoglobin constitutes approximately one-third of the volume of RBCs and accounts for 90-95% of their dry weight. The hemoglobin concentration in the blood is measured in grams per deciliter (g/dL). In a healthy individual, the normal range of hemoglobin levels falls between 12 and 20 g/dL. It is worth noting that males generally exhibit higher Hb levels compared to females. For males, the typical range is 13.5 to 17.5 g/dL, while for females, it is 12 to 15.5 g/dL. In 1959, Max Perutz made a significant discovery by unraveling the structure of hemoglobin. His findings revealed that hemoglobin is a tetrameric protein consisting of two subunits: ' α' and 'β.' Each subunit is composed of a polypeptide chain that is intricately linked to a heme prosthetic group.

A. α Subunit

The α subunit of hemoglobin comprises an alpha polypeptide chain containing 141 amino acid residues.

B. β Subunit

The β subunit of hemoglobin comprises a beta polypeptide chain containing 146 amino acid residues.

C. Heme Group

This group includes iron molecules with a prosthetic group attached to each polypeptide chain. The iron is located at the center of the porphyrin ring.

In the quaternary structure, the α and β subunits exhibit a strong interaction. After mild urea treatment, hemoglobin dissociates partially while the αβ dimers remain intact. Hydrophobic interactions, hydrogen bonding, and a few ion pairs or salt bridges primarily hold the subunits together. During infancy, there are two alpha and two gamma chains present in hemoglobin, which are later replaced by beta chains.

Hemoglobin exists in two conformations, namely the R state and the T state. Oxygen has a higher affinity for the R state, while deoxyhemoglobin is primarily found in the T state.

Hemoglobin functions as a respiratory pigment, facilitating the transportation of oxygen in the form of oxyhemoglobin from the lungs to various tissues in the body. Oxygen binding to Hb is characterized by cooperative binding, where binding one oxygen molecule enhances the affinity for subsequent oxygen molecules. The release of oxygen in the lungs and its uptake in the tissues occurs due to the transition between the T state (Tense) with low oxygen affinity and the R state (Relaxed) with high oxygen affinity. Additionally, it also transports a certain amount of carbon dioxide back as carbaminohaemoglobin.

Hemocyanin

In certain crustaceans, a protein called hemocyanin contains copper and has a distinct chemical structure from hemoglobin. When oxygenated, hemocyanin displays a blue color, which fades to colorless when oxygen is absent. Among annelids, some possess chlorocruorin, a green pigment containing iron, while others have hemerythrin, a red pigment also containing iron. Hemocyanins are large proteins that carry oxygen and are present in mollusks and arthropods. The oxygen-binding site in these proteins consists of a pair of copper atoms directly bound to amino acid side chains. These copper-containing hemocyanins are crucial in binding, transporting, and storing dioxygen within numerous invertebrates' blood (hemolymph).

Cytochromes

Cytochromes are a class of respiratory pigments bound to the cell membrane and primarily found in the mitochondria of cells. Their main function is to facilitate the transfer of electrons along the respiratory pathway. Cytochromes can be classified into three main types: A, B, and C, each characterized by specific subtypes. These cytochromes, A, B, and C, are crucial in facilitating electron transfer processes in respiration and photosynthesis. Another significant cytochrome, called Cytochrome P450, contains a heme group and plays a crucial role in detoxification.

Flavoproteins

Flavoproteins are a class of proteins containing a riboflavin derivative, a nucleic acid. Flavoproteins are extensively studied enzymes and can have either FMN (Flavin mononucleotide) or FAD (Flavin adenine dinucleotide) as a prosthetic group or cofactor. There are 90 encoded flavoproteins in the human genome, with approximately 84% requiring FAD, 16% requiring FMN, and 5 proteins requiring both. These flavoproteins are predominantly found in the mitochondria. Among all flavoproteins, 90% are involved in redox reactions, while the remaining 10% function as transferases, lyases, isomerases, or ligases. These proteins play crucial roles in various biological processes, such as combating oxidative stress, facilitating photosynthesis, and aiding in DNA repair.

3. Derived proteins

These proteins are formed through the degradation or transformation of simple and conjugated proteins. They can be classified into:

a. **Primary-derived proteins** The size of protein molecules remains largely unchanged in these protein derivatives. Examples include proteans, coagulated proteins, and metaproteins.

b. **Secondary-derived proteins** When proteins undergo excessive modifications in their structure and properties, they are referred to as secondary-derived proteins. Examples include proteoses or albumoses, peptones, and peptides.

Properties of Proteins

Color and Flavor

Proteins are typically colorless and often lack taste. They possess a uniform and crystalline structure.

Shape and Size

Proteins exhibit a wide range of shapes, encompassing both simple spherical formations and elongated fibrillar structures.

Molecular weight

The average molecular weight of an amino acid is 110 Daltons (Da). To provide a larger unit of measurement, a kilodalton (kDa) is defined as 1,000 Daltons. When determining the molecular weight of a protein, one must multiply the number of atoms of each element by the atomic weight of that element.

Typically, the molecular weight of a protein falls within the range of 10,000 to 100,000 Daltons. For instance, a protein with a mass of 63 kDa would have a molecular weight of 63,000 grams per mole. On the other hand, a moderately sized protein consisting of only 300 amino acids would have a molecular weight of 33,000 g/mol. It is worth noting that very large proteins can possess molecular weights as high as 1,000,000 g/mol. In terms of specific examples, the molecular weight of proteins can vary significantly. For instance, Cytochrome C has a molecular weight of 12,400 Da, while Ferritin horse has a much higher molecular weight of 450,000 Da.

Colloidal nature

The large size of proteins contributes to their colloidal nature, resulting in various distinctive properties. These biomolecules exhibit significantly slow diffusion rates and can cause visible turbidity in solution through light-scattering, commonly called the Tyndall effect.

Denaturation

Denaturation is the term used to describe the transformation of a protein's properties, leading to the deprivation of its biological activity. Often, denaturation is accompanied by coagulation, wherein the denatured protein molecules aggregate and settle out of the solution.

Protein can undergo denaturation when exposed to certain agents, such as heat and urea. These agents induce the unfolding of the polypeptide chains without cleaving the peptide bonds via hydrolysis. Denaturing agents primarily disturb proteins' secondary and tertiary structures while preserving the primary structure. When a denatured protein is capable of restoring its initial conformation upon removal of the denaturing agent, this phenomenon is known as renaturation. It is important to note that renaturation is only possible if the protein's primary structure remains unchanged. Denaturing agents can be divided into two categories: physical agents and chemical agents. Physical agents comprise heat, radiation, and alterations in pH. Conversely, chemical agents consist of urea solution, which facilitates the formation of new hydrogen bonds within the protein, along with organic solvents and detergents. These agents can disrupt the structural integrity of proteins, leading to denaturation.

Coagulation

When proteins are exposed to heat and undergo denaturation, they undergo a coagulation process, resulting in the formation of insoluble aggregates known as coagulum. It is important to note that not all proteins exhibit heat coagulation; only certain types, such as albumins and globulins, possess this property.

Isoelectric point

The isoelectric point, commonly referred to as pI, designates the pH value at which the amino acid's overall charge is zero, with an equal number of positive and negative charges. This specific pH condition causes proteins to remain stationary when exposed to an electric field, as they do not migrate toward either the anode or cathode. Consequently, this property is harnessed in the process of protein isolation.

Ion binding capacity

Proteins possess the ability to bind with both cations and anions, forming salts, which are determined by their overall charge.

Optical activity

Protein solutions exhibit a phenomenon known as optical activity, whereby they cause the plane of polarized light to rotate mostly in a counterclockwise direction, commonly referred to as levorotation, indicating their levorotatory nature. However, some proteins can rotate the plane-polarized light in a clockwise direction, depicting their dextrorotatory nature.

Solubility

Protein solubility is significantly affected by pH levels. At the isoelectric point, solubility is at its minimum but gradually increases as the acidity or alkalinity rises. This phenomenon can be attributed to the fact that when protein molecules exist as cations or anions, the repulsive forces between ions become stronger due to the presence of excess charges of the same sign. Consequently, the solubility of these molecules surpasses that of the isoelectric state.

Chemical properties

The chemical properties of proteins are as follows:

Hydrolysis

Various hydrolytic agents can be utilized to break down proteins. Acidic agents, such as concentrated hydrochloric acid (6-12N) at temperatures of 100-110°C for 6 to 20 hours, can induce hydrolysis, resulting in the formation of amino acids. Additionally, alkaline agents like 2N sodium hydroxide can also be employed for protein hydrolysis.

Color reaction with Biuret reagent

The addition of an alkaline CuSO4 reagent to a protein solution initiates a reaction in which the peptide bonds within the protein interact with copper ions, resulting in the creation of a Biuret complex with a violet hue. The intensity of the violet color is dependent on the number of peptide bonds present in the protein. This reaction is extensively employed as a qualitative test for protein detection and a quantitative test for protein estimation in biological materials.

Reaction with alkali

The reaction involving the COOH group of proteins with alkali results in the formation of salts.

Reaction with mineral acids

The treatment of mineral acids, like HCl, with either free amino acids or proteins results in the formation of acid salts.

Reaction with formaldehyde

The interaction with formaldehyde gives rise to the production of hydroxy-methyl derivatives.

Benzaldehyde reaction

The reaction with benzaldehyde results in the formation of Schiff's bases.

Reaction with phenyl isocyanate

When proteins react with phenyl isocyanate, they produce hydantoic acid, which can be further converted into hydantoin.

Protein Structure

Due to its complexity as a macromolecule, protein is characterized by four distinct levels of structural organization. These levels, namely primary, secondary, tertiary, and quarternary, serve as a framework for describing the arrangement of protein molecules. Primary, secondary, and tertiary structures can be observed in proteins that consist of a single polypeptide chain. Conversely, the quarternary structure involves the interactions between polypeptides within a protein molecule that comprises multiple chains.

1. Primary structure

The protein's primary structure is relatively simple, comprising one or more amino acid units. It reveals the sequence and quality of amino acids that form the polypeptide chain. The presence of a partial double bond between carbon and nitrogen in the amide bond plays a crucial role in maintaining the stability of the peptide bond. A resonance effect further enhances this stability, where the lone pair of electrons from the nitrogen is donated to the carbonyl group. Through the resonance mechanism, electrons can be shared among multiple atoms, leading to the formation of a resonance structure that is more stable than the original structure. While this increased stability contributes to the strength of the bond, it also restricts the rotation of the amide bond due to the presence of a partial double bond. As a result, the peptide bond adopts a planar configuration with limited movement around the C-N bond, while the single bonds adjacent to the C-N bond experience a greater degree of rotational motion.

2. Secondary structure

The secondary structure of proteins is determined by the spatial arrangement of amino acids that are close to each other in the amino acid sequence. The primary structure, which is initially linear and unfolded, undergoes a conformational change and adopts a helical shape to form the secondary structure. This folding process is primarily influenced by the presence of hydrogen bonds, which can occur within the same molecule (intramolecular) or between different molecules (intermolecular). The interaction of neighboring amino acids through folding and hydrogen bonding results in forming a helix, which exhibits a rigid and tubular structure. Proteins showcase two distinct types of secondary structures, namely α-helix and β-pleated sheet, which are classified based on the specific nature of the hydrogen bonding.

Alpha helix

The alpha-helix is the most prevalent secondary structure found in proteins. Its structure was initially proposed by Linus Pauling and later confirmed through the analysis of the three-dimensional structure of myoglobin using X-ray crystallography. A hydrogen bond is established between the N–H group and the C=O group of adjacent amino acids in the alpha-helix. It is important to note that the alkyl groups within the alpha-helix chain do not participate in these hydrogen bonds. Still, they do contribute to maintaining the overall structure of the helix. Each complete turn of the alpha helix consists of approximately 3.6 amino acid residues. The stability of the alpha helix is primarily attributed to the hydrogen bonds formed between the CO and NH groups of the main chain.

β-pleated sheet

The beta-pleated sheet structure is formed by arranging several polypeptide chains in a parallel orientation, where hydrogen bonds are established between the >C = O and N-H groups of neighboring chains. Within a single polypeptide chain, the R groups of the amino acids are positioned alternatively above and below the plane of the sheet in the beta-pleated structure. The formation of beta-pleated sheets involves both intermolecular and intramolecular hydrogen bonding, with the C=O and N-H groups playing a crucial role in the process. This arrangement is typically stabilized by external interactions involving polar hydrophilic hydrogen and ionic bonds. At the same time, its stability is contributed to by internal hydrophobic interactions among nonpolar amino acid side chains.

3. Tertiary structure

The tertiary structure of a protein molecule is characterized by its three-dimensional conformation, which results from the folding of its secondary structure in specific patterns. This arrangement is typically stabilized by external interactions involving polar hydrophilic hydrogen bonds and ionic bonds, while internal hydrophobic interactions between nonpolar amino acid side chains contribute to its stability.

4. Quarternary structure

The quaternary structure describes the three-dimensional arrangement of protein subunits in proteins containing multiple identical or different polypeptide chains. Noncovalent forces hold these subunits together, which arise from the complementary hydrophobic and hydrophilic regions on their surfaces. These forces facilitate rapid conformational changes in the polypeptide chains, thereby affecting the biological activity of the proteins.

Creatine phosphokinase is an example of a protein that consists of two monomers, forming a dimer. In contrast, hemoglobin is a protein composed of four monomers, making it a tetramer.

Functions of proteins

Proteins play vital roles in various biological activities. Some of the important functions are mentioned below:

- They act as catalysts, speeding up chemical reactions in metabolic pathways.
- Fibrous proteins hold tissues together, like collagen in connective tissues.
- The presence of contractile proteins, such as actin and myosin, is responsible for the capacity of motion and flexibility in organisms.
- Nucleoproteins carry genetic information, influencing traits.
- Proteins also regulate the transport of compounds in and out of cells.
- Hormones made of proteins control growth and physiological functions.
- Soluble proteins in blood plasma can treat shock from severe injuries.
- Interferons are regulatory glycoproteins produced in response to infections.
- Human peptides called defensins have antibiotic properties.
- Thrombin and fibrinogen are examples of proteins that play a role in blood coagulation.

Summary

- Amino acids are the building blocks of proteins and play an essential role in various biological activities.
- Peptides comprise a concise sequence of 2 to 50 amino acids linked together through a covalent bond formed by a condensation reaction.
- A polypeptide is a sequence of amino acids that are connected by peptide bonds.
- Denaturation is a transformative process that alters a protein's molecular composition, disrupting the bonds responsible for its inherent structure.
- The isoelectric point (pI) refers to the pH value at which an amino acid exists as the zwitterion.
- Protein molecules can display up to four levels of structure.
- A zwitterion is an ion that comprises two functional groups. In simpler words, it is an ion that possesses both positive and negative electrical

charges. Hence, zwitterions are primarily electrically neutral, with the net formal charge usually being zero.

Long Type Questions

1. Describe proteins along with their classification?
2. Write a short note on peptides and polypeptides?
3. Write a short note on the following terms:
   a. Keratin
   b. Elastin
   c. Collagen
   d. Albumin
   e. Globulin
   f. Glutelin
   g. Histones
4. Write a short note on the following terms:
   a. Hemoglobin
   b. Hemocyanin
   c. Cytochromes
   d. Flavoproteins
5. Mention the physical and chemical properties of the proteins?.
6. Describe the structure of protein?.
7. Mention the functions of the proteins?.

Fill in the Blanks

1. Proteins are large molecules composed of ____________________
2. Proteins consist of amino acids that are connected through__________
3. Peptide bonds are formed by the linkage between the __________ group of one amino acid and the______________ group of another amino acid.
4. Dipeptides are peptides composed of_____________ amino acids, while polypeptides are those with more than____________ amino acids.
5. Keratin is a ______________ protein and belongs to the____________ family.
6. Albumins are commonly found in ____________and belong to the family of _________ proteins.
7. There are __________main types of _________________, categorized based on their density and the ratio of proteins to lipids.

6

Nucleic Acids

Nucleic acids are complex, chain-like polymers consisting of nucleotides as their monomers. Due to this characteristic, they are often referred to as polynucleotides. The largest molecules present in the biological system are nucleic acids. The two main types of nucleic acids are DNA (deoxyribonucleic acid) and RNA (ribonucleic acid). The bulk of DNA is confined to the nucleus and some cell organelles like mitochondria and chloroplast. RNA is found in both the nucleus and cytoplasm. In 1869, Johannes Friedrich Miescher isolated a substance from the nuclei of the pus cell (leukocytes). The extracted material was named nuclein, which contained an unusually large amount of phosphorus; in addition to proteins, and devoid of sulfur. Richard Altman later named the nuclein as nucleic acid. These organic molecules are found in living cells and play a crucial role in transmitting genetic information from one generation to another.

Nucleotides

Nucleotides serve as the fundamental constituents of nucleic acids. Nucleotides have flat planar and heterocyclic origin. Nucleotides are small molecules composed of three components:

1. Nitrogenous base,
2. Pentose sugar and
3. Phosphate group.

Nucleotides play many roles in the cell, synthesizing vitamin B complex, signal transduction, lipid synthesis, glycogen synthesis, etc.

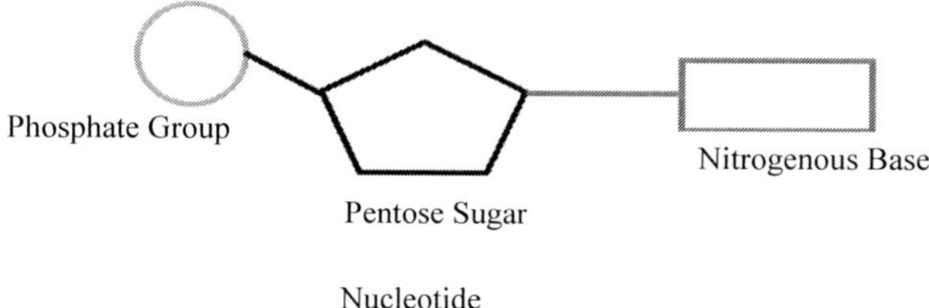

Nucleotide

Nitrogenous bases

Nitrogenous bases are a type of organic molecules that possess a ring structure comprising carbon and nitrogen atoms. Specifically, one of the nitrogen atoms in the ring structure possesses a lone pair of electrons, which enables it to function as a Lewis base. This means that it can donate a pair of electrons during a chemical reaction. There are two distinct forms of nitrogenous bases in both DNA and RNA, i.e., purines and pyrimidines.

Nucleotide = Nitrogenous base + sugar + phosphate

Or

Nucleotide = Nucleoside + phosphate group

Purines

Purines are nitrogenous bases that consist of two fused rings, i.e., a five-membered and a six-membered. The two important purine bases found in the DNA and RNA are adenine and guanine, represented by the letters A and G. Adenine (A) has an amino group at position 6 of its six-membered ring, also known as 6-aminopurine. Similarly, guanine (G) has amino nitrogen at position 2 and oxygen at position 6 of the six-membered ring and hence is also known as 2-amino-6-oxypurine.

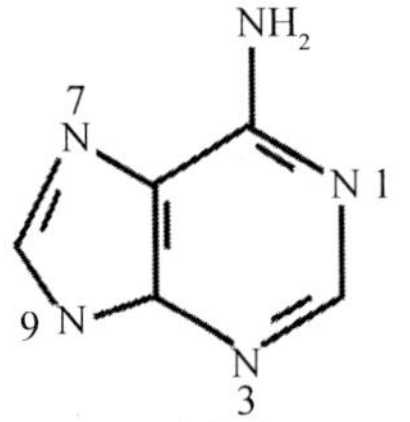

Adenine

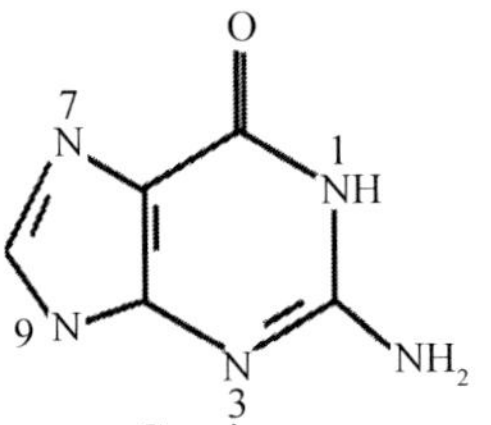

Guanine

Pyrimidine Bases

Pyrimidines are nitrogenous bases that contain a single, six-membered ring. Three pyrimidine bases are known to exist in nucleic acids, i.e., thymine (T), cytosine (C), and uracil (U). Thymine is found only in DNA, and uracil is found only in RNA.

Cytosine has amino and oxy group at 2 and 4 positions, respectively, and hence is also known as 2-oxy-4-aminopyrimidine. Similarly, thymine has two oxy groups (position 2, 4) and one methyl group at position 5 and hence is also known as 2,4-dioxy-5-methylpyrimidine. Uracil is also known as 2,4-dioxypyrimidine as it contains two oxy groups at locations 2 and 4; it is similar to thymine, except it lacks a methyl group. Cytosine is found in DNA and RNA, whereas thymine is found only in DNA and uracil in RNA only.

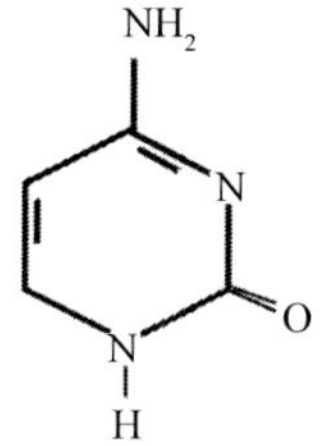

NH2 H3C O H O H N N N O O O N N N H H H

Cytosine 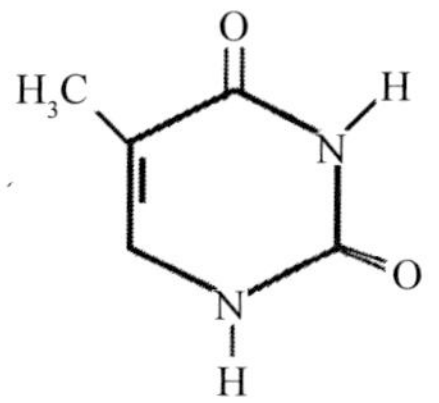 Thymine 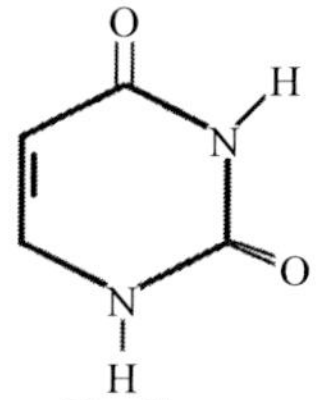 Uracil

Sugars

The pentose sugars are found in DNA and RNA. Deoxyribose in DNA and ribose in RNA are sugars that differ slightly from each other. Both sugars are present in the β-anomeric form. The pentose sugar consists of five carbon atoms in its structure and is in the furanose form. Each carbon atom within the sugar molecule is designated with a unique number, namely 1′, 2′, 3′, 4′, and 5 ′ (pronounced as "one prime").

Deoxyribose

Deoxyribose or 2-deoxyribose is an aldopentose sugar found in DNA that lacks the OH group at position 2 (C-2). The credit for the discovery of deoxyribose goes to Phoebus Levene, who made these significant findings in 1929. Due to this one structural difference, deoxyribose is more stable as compared to ribose sugar. The chemical formula of deoxyribose is $C_5H_{10}O_4$. The molar mass of deoxyribose is 134.13 g/mol.

Ribose

Ribose is also an aldopentose sugar found in RNA and has the chemical formula of $C_5H_{10}O_5$. In 1891, Emil Fischer and Oskar Piloty were credited with the discovery of ribose. The molar mass of ribose is 150.13 g/mol.

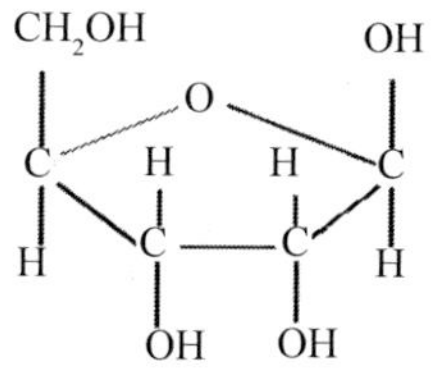

CH2OH OH CH2OH OH O O C H H C C H H C C C C C H H H H OH OH OH OH

2- Deoxyribose 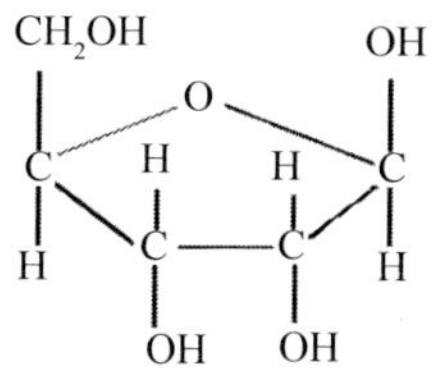 Ribose

Phosphate

A phosphate ion (PO_4^{3-}) or a phosphoric acid molecule can symbolize the phosphate unit, imparting a negative charge to the nucleic acids. The presence of nucleotides can be characterized by the presence of either a single phosphate group or a chain of up to three phosphate groups, which are attached to the 5' carbon of the sugar. However, upon elongation of the DNA or RNA chain, the newly added nucleoside triphosphate will lose two phosphate groups and hence is added only in the form of nucleoside monophosphate by forming a phosphodiester bond.

Nucleosides

Nucleosides are solely composed of pentose sugar and nitrogenous bases and lack a phosphate group. In nucleoside, a pentose sugar, is either ribose or deoxyribose that is chemically bonded (β- glycosidic bond) to a nitrogenous base, which can be a purine or pyrimidine.

Nucleoside = Nitrogenous base + Sugar

Nomenclature of bases

When a deoxyribose sugar is linked with a nitrogenous base, it is referred to as deoxyribonucleoside. A deoxyribonucleoside that possesses one phosphate moiety is known as deoxynucleoside monophosphate,whereas those possessing two and three phosphate groups are referred to as deoxynucleoside diphosphate and deoxynucleoside triphosphate.

Similarly, when a ribose sugar is attached to a nitrogenous base, it is known as ribonucleoside. On the other hand, when a single phosphate group is attached to the ribonucleoside moiety, it is known as ribonucleoside monophosphate. Moreover, when two and three phosphate groups are attached to the ribonucleoside monophosphate, the structures are referred to as ribonucleoside diphosphate and ribonucleoside triphosphate, respectively.

Base	**Nucleosides**	**Nucleotides**
	DNA	
Adenine	Deoxyadenosine	Deoxyadenylate Deoxyadenosine-5'- monophosphate (dAMP)
Guanine	Deoxyguanosine	Deoxyguanylate Deoxyguanosine-5'- monophosphate (dGMP)
Cytosine	Deoxycytidine	Deoxycytidylate Deoxycytidine-5'- monophosphate (dCMP)
Thymine	Deoxythymidine	Deoxythymidylate Deoxythymidine-5'- monophosphate (dTMP)

Base	Nucleosides	Nucleotides
	RNA	
Adenine	Adenosine	Adenylate Adenosine-5'- monophosphate (AMP)
Guanine	Guanosine	Guanylate Guanosine-5'- monophosphate (GMP)
Cytosine	Cytidine	Cytidylate Cytidine-5'- monophosphate (CMP)
Uracil	Uridine	Uridylate Uridine-5'- monophosphate (UMP)

Deoxyribonucleic acid

Deoxyribonucleic acid, or DNA, is a polynucleotide composed of nucleotides arranged in a long chain. Each nucleotide comprises a nitrogenous base linked to deoxyribose sugar through a glycosidic bond and a phosphate group. The structure of DNA is characterized by a phosphodiester backbone composed of alternating sugars and phosphates, with the nitrogenous bases occupying the interior. DNA serves as the hereditary material in most living organisms, ranging from single-celled bacteria to multicellular mammals (both prokaryotic and eukaryotic cells). Its discovery and identification can be attributed to the Swiss biologist Johannes Friedrich Miescher, who made this breakthrough during his study on white blood cells in 1869. DNA transmits genetic instructions or hereditary materials from parents to offspring. In eukaryotic cells, nuclear DNA is located within the nucleus. It encodes most of an organism's genomes, whereas extranuclear DNA is present in mitochondria (mitochondrial DNA), and chloroplast (plastid DNA) is responsible for the remaining function.

Structure of DNA

James Watson, a renowned American biologist, and Francis Crick, a distinguished British physicist, collaborated in the early 1950s to create their famous model of the DNA double helix. Instead of conducting fresh experiments in the laboratory, Watson and Crick primarily relied on collecting and analyzing existing data, ingeniously combining them in novel and insightful ways. A significant portion of their crucial insights into the structure of DNA was derived from the contributions of Rosalind Franklin (X-ray diffraction), a chemist working under the guidance of physicist Maurice Wilkins. In 1962, James Watson, Francis Crick, and Maurice Wilkins were jointly granted the Nobel Prize in Medicine for their groundbreaking discovery. Unfortunately, Franklin had already passed away then, and Nobel Prizes were not awarded posthumously.

Watson and Crick model

In 1953, James Watson and Francis Crick proposed the idea of the double helix structure of DNA with a specific pairing of nucleotides. The general features of this double helical model are as follows.

1. The structure of a DNA molecule consists of two biopolymer nucleotide strands that twist with each other, forming a double helix structure resembling a twisted ladder.
2. The nucleotide strands are made up of monomeric units called deoxyribonucleotides. These deoxyribonucleotides are connected through phosphodiester bonds.
3. The deoxyribonucleotides contain nitrogenous bases (adenine, cytosine, thymine, and guanine), deoxyribose sugar, and a phosphate group.
4. One end of each strand is referred to as the 5′end, characterized by a phosphate group, while the other end is known as the 3′end, featuring a hydroxyl group.
5. These two polynucleotide chains/strands run in opposite directions, i.e., they are antiparallel to each other, wherein one strand runs in a 5′to 3′direction, while the other strand runs in a 3′ to 5′direction.
6. The two strands are connected through hydrogen bonds and exhibit a complementary base pairing nature to one another. Through two hydrogen bond interactions, adenine (A) consistently base pairs with thymine (T), while guanine (G) forms a stable base pair with cytosine (C) through three hydrogen bonds. This principle, known as the rule of the base pair, underscores the complementary nature of these nucleotides.
7. Chargaff's assumptions play a crucial role in determining the structure of double-helical DNA. According to these assumptions, the base pairs in DNA follow a specific pattern: the amount of adenine (A) is equal to thymine (T), and guanine (G) is equal to cytosine (C). This means that the sum of purines (A+G) always equals the sum of pyrimidines (C+T).
8. The double helix has a diameter of 20Å, and its helical repeats is 34Å, which precisely corresponds to 10 base pairs.
9. The presence of a double-helical configuration in DNA leads to the formation of two asymmetrical grooves within the molecule. When the backbones of a DNA molecule are far apart, a wider groove known as the major groove is formed. Conversely, a narrower groove called the minor groove is formed when the backbones are close together. These grooves play a crucial role in revealing the edges of the bases, enabling one to decipher the base sequence of a specific DNA molecule, and also play a role in gene regulation.

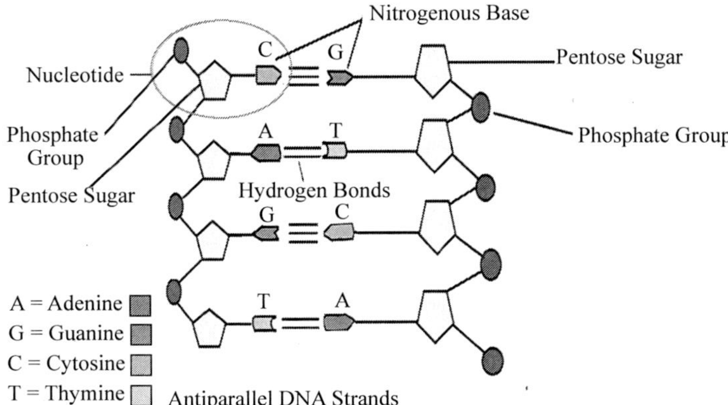

Forms of DNA

Various forms of DNA can arise through natural or synthetic means by altering environmental factors such as humidity, salt levels, pH, and temperature. X-ray diffraction analysis at atomic resolution levels has discovered a wide range of DNA configurations in DNA crystals, including B, A, C, D, E, and Z. Some of the significant forms of DNA are mentioned below.

1. B-DNA

The B-DNA structure first elucidated by Watson and Crick is the most prevalent form of DNA. The majority of cells typically exhibit this type of DNA under normal physiological conditions, characterized by 92% relative humidity and low ionic strength. The B-DNA form exhibits a right-handed coiling pattern. The B-form helix has a diameter of 20Å, and each turn of the helix consists of 10 base pairs perpendicular to the helical axis. The bases in the B-form exhibit complementary base pairing, with A always pairing with T and G always pairing with C. The B-form structure exhibits plectonemic coiling, where the two strands of the helix intertwine. This structure is considered "Asymmetric" because the major and minor grooves alternate. The major groove is wide and shallow, measuring 12Å in width, while the minor groove is narrow, measuring 6Å in width. The sugar pucker is in the "C2" endoform in the B-type structure.

2. A-DNA

Rosalind Franklin made a significant discovery in her pursuit of unraveling DNA structure through X-ray diffraction. A form of DNA is observed when DNA is dehydrated, serving as a protective mechanism during extreme conditions like desiccation. Additionally, when proteins bind to DNA, it also adopts the A-form. In this form, the DNA helix has a diameter of 23Å and completes a full helical turn with 11 bases. Comparatively, the A-form is shorter and more compact than the B-form while maintaining its right-handed helical nature. The length of the helix is reduced to 28.15Å. Notably, the base pairs exhibit a tilting of 20.2Å from the helix axis, each measuring approximately 2.56Å in axial size. The A-form showcases a major groove that is narrower and wider in its structure, while the minor groove is wide and shallow. In the C3 endoconformation, the presence of the sugar pucker ring is observed.

3. Z-DNA

Z-DNA is a distinct variation of DNA that differs from B-DNA and A-DNA as it possesses a left-handed double-helical structure and is formed under high salt concentration (ionic) and purine-pyrimidine-rich conditions. The Z-DNA was first discovered by scientists Rich, Northeim, and Wang in 1984. The coiling pattern of Z-DNA exhibits a unique left-handed and zig-zag pattern. Compared to B-DNA, Z-DNA exhibits a longer and thinner conformation, featuring an asymmetric structural configuration that eliminates internal space. Z-DNA is formed through the alternating stretching of purines and pyrimidine bases. The Z-form helix has a diameter of 18Å, and the tilt of base pairs measures 7Å, which is lower than the B and A-form of DNA. In Z-form DNA, a single helical turn comprises 12 base pairs that are perpendicular to the helical axis. Additionally, Z-form DNA exhibits a zig-zag plectonemic coiling pattern. Similar to B-DNA, it consists of antiparallel strands. In the Z-type conformation, the sugar pucker or deoxyribose ring is located in the "C3" endoform for purine bases and the "C2" endoform for pyrimidine bases. The major groove of Z-type DNA is flat, whereas the minor groove is narrow and deep.

4. C-form

C-DNA, also known as circular DNA, is present in limited quantities and is typically found in an environment with 66% relative humidity and low ionic strength. C-DNA, unlike Z-DNA, possesses a right-handed twist in its helical conformation, with a helix pitch of 30.97Å. Approximately 9.33 base pairs are encompassed in each complete turn of the helix. The helix diameter of C-DNA measures 19.0Å. Regarding structural features, C-DNA exhibits a rotation of 38.6 degrees per base pair, accompanied by a base pair tilt of 7.8 degrees. Furthermore, it is worth mentioning that C-DNA is smaller and has a different shape than B-DNA and A-DNA.

5. D-form

D-DNA is a relatively uncommon occurrence within cells, as it is found in very low quantities. The D-DNA structure is distinguished by its right-handed helical twist, referred to as the poly (dA-dT) and poly (dG-dC) conformation.

Within this D-form, there are 8 base pairs present in each complete helical turn, and these base pairs are positioned in a retrograde manner compared to the regular DNA helix. Furthermore, the two strands of D-DNA run in opposite directions, resulting in an antiparallel arrangement. The base pair tilt of D-DNA demonstrates a negative tilt of approximately -16.7 degrees, accompanied by an axial rise of 3.03Å per base pair.

6. E-form

E-DNA, a form of DNA that is infrequently observed in cells, displays distinct features, including an extended or eccentric DNA structure. The helix of E-DNA is notably elongated, with the bases positioned perpendicular to the helical axis. E-DNA is characterized by a deep major groove and a shallow minor groove. Following an in-depth X-ray crystallographic investigation, it has been determined that E-DNA occupies an intermediate position between B-DNA and A-DNA.

Chargaff's rule

Erwin Chargaff, a distinguished Biochemist, made a groundbreaking discovery referred to as Chargaff's Rule. This rule elucidates that the quantity of nitrogenous bases within double-stranded DNA remains consistently balanced. Specifically, the amount of adenine (A) is equal to the amount of thymine (T), and the amount of guanine (G) is equivalent to the amount of cytosine (C). It is noteworthy that the sum of purines (adenine and guanine) is equal to the sum of pyrimidines (thymine and cytosine). This equilibrium of nitrogenous bases ensures that the DNA within a cell maintains an equal proportion of purine and pyrimidine bases, thereby upholding a 1:1 ratio.

Ribonucleic acid (RNA)

Ribonucleic acid (RNA) is a vital biomolecule found in the majority of living organisms, including viruses. It is composed of nucleotides, which consist of ribose sugars attached to nitrogenous bases through a glycosidic linkage and phosphate groups. The nitrogenous bases in RNA include adenine, guanine, cytosine, and uracil. While RNA is closely related to DNA, it differs in its sugar component, containing ribose instead of deoxyribose, and it replaces thymine (T) with uracil (U) as a nitrogenous base. RNA is single-stranded mainly, with a backbone made up of alternating phosphate groups and ribose sugars. However, it is also capable of forming double-stranded RNA (dsRNA), and through complementary base pairing, a single RNA molecule can form intrastrand double helices, as seen in tRNA. It plays a crucial role in different biological processes, either by acting as a genetic material (viruses) or directly performing other functions itself (non-coding RNA) and by serving as a template for protein production (coding RNA).

Forms of RNA

There are a number of different forms of RNA known to date, like messenger RNA (mRNA), ribosomal RNA (rRNA), transfer RNA (tRNA), small nuclear RNA (snRNA), and small nucleolar RNA (snoRNA).

Messenger RNA

mRNA is a specific type of coding RNA molecule. It serves as a single-stranded template for protein synthesis carried out by ribosomes. Jacob and Monad coined the term "messenger RNA." This molecule is composed of nucleotides and forms a long polymeric structure. mRNA consists of a single strand of nucleotides connected by phosphodiester bonds. It contains four nitrogenous bases: adenine, guanine, cytosine, and uracil. The mature mRNA molecule comprises several distinct regions, including the 5′ cap, 5′ untranslated region (UTR), coding region, 3′ UTR, and poly(A) tail. When a coding region of mRNA encodes a single protein, it is referred to as monocistronic, which is the case for most eukaryotic mRNAs except for the mitochondrial mRNA, which exhibits a polycistronic nature. In contrast, prokaryotic mRNAs often code for multiple proteins and are termed polycistronic. Proteins with similar functions are commonly associated and governed by a singular regulatory region, encompassing both promoter and operator regions.

Ribosomal RNA

rRNA is a type of non-coding RNA molecule. During the translation of mRNA into proteins, rRNA plays a fundamental role in all living cells. rRNA stands as the most prevalent form of RNA within the cells of all living organisms. Ribosomal RNA constitutes a significant portion, approximately 60-80%, of the ribosome's overall weight. Its presence is indispensable for the proper functioning of the ribosome, as it actively participates in all of its essential activities, including mRNA binding, tRNA attraction, and the catalysis of peptide bond formation between amino acids. Within the cytoplasm of a cell, the ribosomal RNA (rRNA) serves as a vital component of the ribosome. Each ribosome is organized into two subunits: the large subunit (LSU) and the small subunit (SSU). Each subunit consists of different types of RNAs. These subunits are sometimes referred to by their size-sedimentation measurements, indicated by a number followed by an "S" suffix. In prokaryotes, the LSU and SSU are referred to as the 50S and 30S subunits, respectively. The 50S subunit is composed of 23S rRNA and 5S rRNA, while the 30S subunit includes 16S rRNA. In eukaryotes, they are slightly larger, with the LSU and SSU being called the 60S and 40S subunits, respectively. The 60S contains 28S rRNA, 5.8S rRNA, and 5S rRNA, whereas the 40S possesses 18S rRNA.

Transfer RNA

tRNA, also known as adaptor RNA, is an RNA molecule that has a crucial role during the process of protein synthesis. There are more than 20 different tRNA molecules consisting of 75-95 nucleotides in size. It acts as a bridge between the messenger RNA (mRNA) and the developing chain of amino acids that form a protein. The name "transfer RNA" comes from its responsibility of transferring amino acids to the ribosome. It can also form secondary (Cloverleaf) and tertiary structures (L-shaped).

Clover Leaf Model of tRNA

The clover leaf model of tRNA is a type of secondary structure formed due to hydrogen bonding between base pairs. It consists of four stems and three loops. All four stems are short double helices with loops at the chain's extremities containing seven or eight bases. The free 3' and 5' ends of the chain are located on the unlooped stem, leading to the formation of the acceptor arm. The three loops are named as TψC loop, anticodon loop, and D loop. The TψC loop has ribothymindine and pseudouridine, whereas the D loop or DHU loop always contains dihydrouridine. An additional loop containing a highly variable number of bases, present in many tRNAs, is called a variable loop or extra arm. Anticodon is the part of the tRNA molecule that interacts with the messenger RNA. Three nucleotides present on one arm of tRNA form the anticodon. These three bases form hydrogen bonds with other nucleotides, specifically with a group of three base sequences on mRNA called codons, which match the exact amino acid that the tRNA molecule contains.

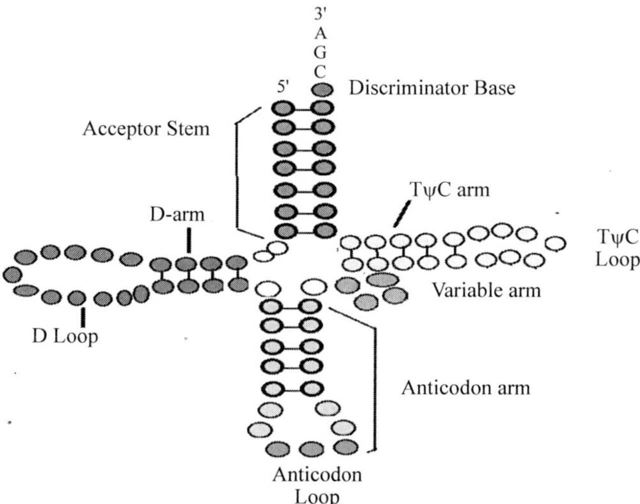

Cloverleaf model of tRNA

Small nuclear RNA (snRNA)

A subset known as snRNA exists among the various RNA molecules involved in transcript slicing. These snRNAs are characterized by their relatively small size, with an average length of approximately 150 nt. In the context of transcript slicing, snRNAs play a crucial role by forming ribonucleoprotein particles (snRNPs) in collaboration with specific proteins. These snRNPs and other proteins assemble into a large complex that associates with unspliced primary RNA transcripts, thereby regulating the splicing process. Additionally, snRNPs have been found to be closely associated with a specific group of proteins and complexes referred to as small nuclear ribonucleoproteins.

Small nucleolar RNA (snoRNA)

Small nucleolar RNAs (snoRNAs) are a class of short, non-protein-coding RNAs responsible for guiding chemical modifications of other RNAs. The intron-encoded ncRNAs represent the largest groups of small RNAs, ranging from 60 to 170 nucleotides in length. They are predominantly found in the nucleolus and are crucial in modifying rRNA. Found in all eukaryotes and a subset of archaea, snoRNAs primarily function within the nucleolus, the most extensive substructure of the nucleus. Their primary role involves modifying and processing ribosomal RNA, but they also serve as molecular chaperones and guide pre-rRNA processing, contributing to various aspects of ribosome biogenesis.

Summary

- Nucleic acids are complex, chain-like polymers consisting of nucleotides as their monomers.
- Nucleic acids are of two types, i.e., DNA (Deoxyribonucleic acid) and RNA (Ribonucleic acid),
- DNA is confined to the nucleus, whereas RNA is found in both the nucleus and cytoplasm.
- Nucleotides comprise three parts: a nitrogenous base, a five-carbon sugar, and a phosphate group.
- Nucleosides are solely composed of sugar and nitrogenous bases, and they lack a phosphate group.
- Deoxyribose in DNA and ribose in RNA are the five-carbon sugars.
- Two purine bases found in the DNA and RNA, i.e., adenine and guanine.
- Three pyrimidine bases are known to exist, i.e., thymine (T), cytosine (C), and uracil (U).

- DNA exists in different forms like A, B, C, and Z-DNA.
- A-form of DNA is right-handed and has 11 base pairs per turn.
- The B-form of DNA consists of 10 base pairs per turn.
- RNA is a single-stranded nucleic acid with ribose sugar and uracil instead of thymine as one of the pyrimidine bases.
- RNA exists in different forms, such as rRNA, tRNA, mRNA, Circular RNA, and LncRNA.

Long Type Questions

1. What are nucleotides? Give the chemical composition of generalized nucleotides.
2. Describe the Watson-Crick model of DNA structure.
3. Write short notes on purines and pyrimidines.
4. Describe briefly different forms of DNA.
5. What are nucleosides? How do they differ from nucleotides?
6. Give a brief account of different forms of RNA.
7. Describe the cloverleaf model of tRNA.

Fill in the Blanks

1. ________are the monomeric unit of nucleic acids.
2. The pyrimidine base found in RNA but not in DNA is _________.
3. The primary structure of DNA and RNA proceeds in _____________ direction.
4. ________ bond is used to stabilize the double helix of DNA.
5. ____________ is the pyrimidine base found in DNA and absent in RNA.
6. DNA was discovered by __________.

7

Enzymes

Enzymes are biocatalysts that play a crucial role in accelerating chemical reactions without undergoing any changes themselves. Enzymes enhance the rate of a reaction by reducing its activation energy without affecting the equilibrium position of the reaction. These catalysts specifically target molecules called substrates and convert them into different molecules known as products. Most enzymes have a proteinous nature, and some have a ribonucleic acid nature (like ribozymes and peptidyl-transferase), showing high substrate and stereo-specificity. Enzyme catalysis is indispensable for the majority of metabolic processes in cells, as it ensures that these processes occur at a rate that is sufficient to sustain life. The progression of metabolic pathways heavily relies on the catalytic activity of enzymes to facilitate individual steps.

The proteinous enzymes are classified into simple and conjugated enzymes.

1. Simple enzymes: Enzymes are made up of only protein components.
2. Conjugated enzymes: Enzymes comprise both protein (apoenzyme) and non-protein components (cofactor).

Cofactors

Enzymes require the presence of non-proteinous substances called cofactors to carry out their functions effectively. These cofactors are essential for the proper functioning of enzymes. The combination of an enzyme and its cofactor forms a holoenzyme. Three distinct types of cofactors can be found in enzymes.

Prosthetic group

Prosthetic groups are non-peptide compounds that exhibit a strong binding affinity towards proteins or enzymes. These specialized non-polypeptide units play an essential role in facilitating the biological function of specific proteins. Some examples of prosthetic groups are heme, copper, biotin, ubiquinone, etc.

Coenzymes

Coenzymes, which are non-protein organic compounds, play a crucial role in catalyzing reactions by binding with enzymes. Although they are commonly referred to as cofactors, it is important to note that coenzymes have distinct chemical compositions. For apoenzymes to carry out their catalytic functions effectively, they often rely on coenzymes. These coenzymes, such as nicotinamide adenine dinucleotide (NAD), flavin adenine dinucleotide (FAD), etc., are not permanently attached to the enzyme but rather exist independently. Many coenzymes are derived from B-complex vitamins and can act as carriers of electrons during reactions or be transferred between enzymes as functional groups.

Metal ions

Metal ions are non-protein inorganic compounds loosely attached to the enzymes, thereby enabling catalysis. The active site of certain enzymes requires the presence of a metal ion to form coordinate bonds. They are a distinct class of atom compounds that exhibit an electric charge. These ions, referred to as cations mostly, are positively charged and arise from the loss of electrons by metals. Their absence can result in growth disorders, profound malfunction, the initiation of carcinogenesis, or even fatality. Notable examples of metal ions encompass potassium (k^+), sodium (Na^+), calcium (Ca^{2+}), magnesium (Mg^{2+}), zinc (Zn^{2+}), copper (Cu^{2+}), and manganese (Mn^{2+}). Metal ions are indispensable for sustaining life in plants, animals, and humans.

Enzymes may act within or outside the cell and may be intracellular or extracellular.

Intracellular enzymes

These are also referred to as endoenzymes and are enzymes that operate within the cells. These enzymes catalyze a wide range of metabolic reactions within the cell, earning them the name metabolic enzymes. Most enzymes found in plants and animals are classified as intracellular enzymes or endoenzymes.

Extracellular enzymes

Extracellular enzymes, or exoenzymes, are enzymes produced by living cells and facilitate important reactions outside the cell but within its immediate environment. Their primary function is to act as digestive enzymes, catalyzing the breakdown of complex macromolecules into simpler polymers or monomers. They are commonly found in bacteria, fungi, and certain insectivores like Drosera and Nepenthes.

Nomenclature of enzymes

The naming of enzymes often involves adding the suffix '-ase' to the name of the substance they act upon. For example, urease is responsible for the hydrolysis of urea, and fructose-1,6-bisphosphatase breaks down fructose-1,6-bisphosphate. However, not all enzymes follow this naming convention.

Enzymes like trypsin and chymotrypsin have names that do not indicate their substrate. Moreover, some enzymes have multiple names. An internationally agreed enzyme nomenclature system called the Enzyme Commission numbering system has been implemented to establish a standardized system for enzyme names. This system classifies enzymes into six major classes based on the type of reaction they catalyze. Each enzyme is then assigned a unique four-digit classification number.

For instance, the Enzyme Commission (EC) number 3.4.21.4 is assigned to trypsin, wherein.

1. The first numerical value (**3**) represents a **trypsin class**, categorized into the hydrolase class.
2. The second numerical value (**4**) signifies a **sub-class** of trypsin, as it functions as a protease responsible for breaking down peptide bonds.
3. The third numerical value (**21**) indicates its **sub-sub-class** as a serine protease containing a crucial serine residue at its active site.
4. Lastly, the fourth numerical value (**4**) denotes the **order** of trypsin, as it was the fourth enzyme to be classified within this class.

To compare, chymotrypsin is assigned the EC number 3.4.21.1, while elastase is assigned 3.4.21.36.

Classification of enzymes

The International Union of Biochemists (IUB) has classified enzymes into six distinct functional classes based on the specific type of reaction they facilitate as catalysts. These classes include oxidoreductases, transferases, hydrolases, lyases, ligases, and isomerases.

1. Oxidoreductases

Oxidoreductases are a class of enzymes responsible for facilitating the transfer of electrons between molecules. Specifically, these enzymes catalyze the movement of electrons from a molecule known as the reductant or electron donor to another molecule referred to as the oxidant or electron acceptor (involved in oxidation and reduction reactions). Additionally, within the category of this class, dehydrogenase, peroxidases, oxidases, oxygenases, and reductases are also included. In dehydrogenases, the final electron acceptor or donor is NADH or FAD, for example, alcohol dehydrogenase, malate dehydrogenase, etc. In peroxidases, the final electron acceptor is hydrogen peroxide. The final electron acceptor in oxidases is an oxygen molecule like cytochrome oxidase, amino acid oxidase, etc. In contrast, oxygenases involve the addition of oxygen atoms like mono-oxygenase to the substrate.

2. Transferases

Transferases are a class of enzymes that facilitate the transfer of a functional group from one molecule, known as the donor molecule, to another molecule, referred to as the acceptor molecule. They play a crucial role in various biological processes by catalyzing the transfer of specific functional groups. One example of a transferase is acyltransferase, which transfers acyl groups between molecules, thereby contributing to the modification and regulation of various cellular processes. N-acetyltransferase is another type that is essential for a specific metabolic pathway that utilizes tryptophan, an essential amino acid. This transferase transfers an acetyl group to molecules involved in the tryptophan pathway, thereby facilitating the synthesis of essential compounds and metabolites. DNA methyltransferase is another essential transferase that transfers a methyl group to a DNA acceptor molecule. This enzymatic process, known as DNA methylation, involves the addition of a methyl group to the DNA molecule, which can have significant effects on gene expression, regulation, and ensuring the stability and integrity of the genome. The other examples of class transferases are kinases, transaminases, transcarboxylases, etc.

3. Hydrolases

Hydrolases, a category of enzymes, function as biochemical catalysts by utilizing water to cleave chemical bonds. This enzymatic process typically results in the division of larger molecules into smaller entities. Esterases, lipases, phosphatases, glycosidases, peptidases, and nucleosidases are commonly encountered hydrolase enzymes. Esterases specifically target ester bonds present in lipids, whereas phosphatases are responsible for removing phosphate groups from molecules. Similarly, nucleases are a class of enzymes that facilitate the hydrolysis of nucleic acids, breaking them down into their constituent parts. In contrast, glycoside hydrolases catalyze the hydrolysis of glycosidic bonds, leading to the breakdown of complex carbohydrates.

4. Lyases

Lyases are a class of enzymes that facilitate chemical reactions by introducing functional groups (CO_2, NH_3, and H_2O) to double bonds or eliminating functional groups to generate double bonds. Frequently, they are responsible for the formation of double bonds or the incorporation of new ring structures. Decarboxylases, deaminases, dehydratases, synthases, etc., are commonly encountered hydrolase enzymes. Decarboxylase involves the removal of CO_2, deaminase involves the removal of NH_3, and dehydratase removes H_2O.

5. Isomerases

Isomerases are a class of enzymes that play a crucial role in catalyzing the transfer of atoms or groups within a single molecule. Isomerases, epimerases, and mutases belong to this class. Their primary function is facilitating intramolecular rearrangements, thereby converting a molecule from one isomer to another. This process involves the breaking and forming chemical bonds, enabling the transformation of the molecular structure. The enzyme phosphohexose isomerase facilitates the conversion of D-fructose to D-glucose and vice versa, playing a crucial role in the interconversion of these two sugars. Phosphotriose isomerase enzyme facilitates the conversion of glyceraldehyde-3-phosphate into its isomer, dihydroxyacetone phosphate.

6. Ligases

Ligases, also referred to as synthetases, are a class of enzymes that facilitate the formation of a new chemical bond by catalyzing the joining of two components. Notable examples are glutamine synthetase, acyl-CoA synthetase, DNA ligase, etc. DNA ligase, specifically, plays a crucial role in both DNA repair and DNA replication processes. Moreover, this enzyme finds extensive utility in various molecular biology and biotechnology applications. T4 DNA ligase, an enzyme derived from the bacteriophage T4, holds significant importance in the field of recombinant DNA technology as it serves as one of the frequently employed DNA ligases.

Enzyme units

The enzyme units are used to express the catalytic activity of an enzyme. The two main enzyme units are the Katal and Enzyme units. Katal (Kat) is defined as the conversion of one mole of the substrate into a product per second. At the same time, enzyme unit (E.U) is defined as the conversion of micromoles of the substrate into product per minute. The E.U is also known as the international unit (I.U) of an enzyme and is an S.I unit of an enzyme.

Mechanism of enzyme action

The mechanism of enzyme activity is contingent upon the characteristics of both the enzyme and the substrate molecule. This can be comprehended through the following processes:

1. The formation of an enzyme-substrate (ES) complex.
2. The lowering of activation energy.

1. The formation of an enzyme-substrate complex

Enzymes possess a specific region called the active site, which has a defined shape and functional groups that enable them to bind with reactant molecules.

The molecule that binds to the enzyme is referred to as the substrate. When the substrate and enzyme interact, they form an intermediate with a lower activation energy, even in the absence of catalysts. In every enzymatic reaction, the enzyme and the substrate initially come together to create an unstable complex known as the enzyme-substrate (ES) complex. This complex then breaks apart, resulting in the formation of the product (P) and the release of the enzyme (E).

Enzyme (E) + Substrate (S) $\rightleftharpoons$ Enzyme-substrate (ES) complex $\rightarrow$ Enzyme (E) + Product (P)

On the surface of each enzyme, there are active sites to which the substrate molecules attach forming ES complex which ultimately leads to the formation of the product. Once the product is formed, it separates from the enzyme, allowing the enzyme to bind with other substrate molecules and form a new product, and so on.

Upon the formation of the ES complex, new covalent and non-covalent bonds are formed between substrate and enzyme. The phenomenon of ES complex formation depends upon enzyme specificity, which has been explained through the following theories.

i) Lock and key hypothesis

Emil Fischer proposed this hypothesis in 1898, which provides insight into the mechanism of enzyme action. According to this hypothesis, both the enzyme and substrate molecules possess distinct geometric shapes (rigid) that are highly specific. Similar to the lock and key, this hypothesis suggests that the enzyme functions as a 'lock' and the active site of an enzyme functions as a 'keyhole' with specific molecules, such as -COOH and -SH. At the same time, a substrate functions as a 'key,' and only a specific substrate can unlock these enzyme molecules and form ES complexes. This complex undergoes chemical changes, ultimately resulting in the formation of a product. Once the product is formed, it no longer fits into the active site and is released into the surrounding area. Consequently, the active site becomes available again for new substrates. It also explains how enzymes are not consumed in the reaction and can be reused. Furthermore, it aids in understanding the mechanism of competitive inhibition.

ii) Induced fit hypothesis

Daniel Koshland (1958) proposed this theory, which suggests that the active sites on enzyme surfaces are not rigid but relatively flexible in structure. Initially, these active sites do not match with the substrate molecule. However, when a specific substrate molecule approaches the specific enzyme, it triggers a reorganization of the substrate molecule, which makes it complementary to the active site of the enzyme and finally forms an ES complex. According to this theory, the active sites consist of two functional groups: one aids in substrate binding, while the other facilitates the conversion of the substrate into the product.

2. Lowering of activation energy

For a reaction to occur, two molecules must collide with each other having the appropriate orientation and sufficient energy. This energy, known as activation energy, is required to overcome the barrier. Activation energy represents the minimum energy required to initiate a chemical reaction. Enzymes, like catalysts in general, decrease the activation energy necessary for chemical reactions to commence.

Factors affecting enzyme activity

A range of factors, including temperature, concentration of enzyme/substrate, pH, salt concentration, activators, and inhibitors, can influence enzyme catalytic activity.

Enzyme concentration

In the presence of an adequate amount of substrate, the enzyme reaction rate can be enhanced by elevating the concentration of enzymes.

Temperature

Enzymes necessitate an ideal temperature for their functioning. The temperature at which an enzyme exhibits its maximum catalytic activity is referred to as the optimum temperature. Exceeding the optimum temperature can potentially modify the molecular composition of the enzymes, thereby decreasing their activity. Conversely, cold temperatures impede enzyme activity by reducing molecular motion.

Salt concentration

Optimal salt concentration plays a vital role in determining the functionality of enzymes. Any fluctuations in this concentration change the enzyme activity while disrupting the enzyme's structure and function, leading to denaturation.

pH

For enzymes to perform at their best, they require an ideal pH called optimum pH. The optimum pH refers to the pH level at which a compound demonstrates its maximum activity. Any pH level above or below the optimum range has the potential to disrupt the molecular structure of enzymes, thereby decreasing their activity. Generally, the optimal pH range for enzymes falls between 5 and 7.

Activators and inhibitors

Enzyme activators, also called positive modulators, are specific molecules that augment the catalytic activity of an enzyme, thereby enhancing its overall efficiency. Conversely, enzyme inhibitors, also called negative modulators, are molecules that impede or diminish the activity of an enzyme, leading to a reduction in its catalytic potential.

Substrate concentration

The substrate concentration acts as the limiting factor when the enzyme concentration is constant. That is, with an increase in substrate concentration, the rate of enzyme-catalyzed reaction also increases up to a certain concentration. However, when the substrate concentration reaches a very high level, the enzymes become saturated with substrate concentration, and additional substrate concentration fails to enhance the reaction rate.

Michaelis-Menten equation

Scientists Leonor Michaelis and Maud Leonora Menten formulated the Michaelis-Menten equation in the 20th century to explain the correlation between substrate concentration and the rates of enzyme-catalyzed reactions. As a result, it was named after the German biochemist Leonor Michaelis and the Canadian physician Maud Menten. This equation/model was developed to elucidate the variations in the rate of chemical reactions catalyzed by enzymes. The equation establishes a relationship between the velocity/rate of the reaction and the concentration of the substrate in a system where the substrate (S) binds reversibly to the enzyme (E), forming an enzyme-substrate (ES) complex. This complex then undergoes an irreversible reaction to produce a product (P) and regenerate the free enzyme (E). This mechanism can be visually represented as follows:

$$E + S \rightleftharpoons ES \rightarrow E + P$$

The Michaelis-Menten equation for this system is:

$$V = \frac{V_{max}[S]}{K_M + [S]}$$

The Michaelis-Menten kinetics equation comprises two essential components.

Maximum reaction rate (Vmax) occurs when the substrate occupies all substrate-binding sites in an enzyme.

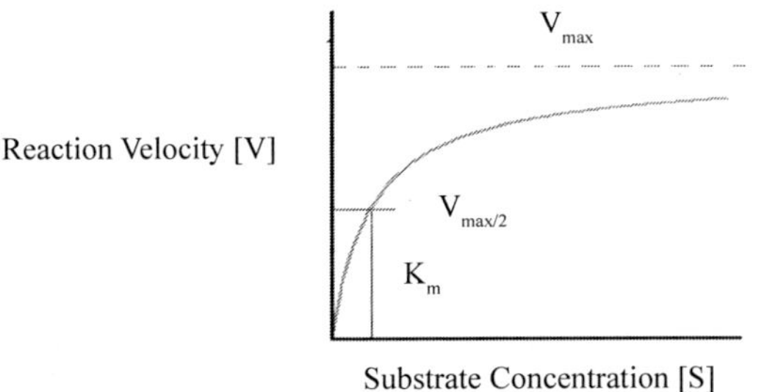

Michaelis-Menten constant or Michaelis constant (K_m) is the concentration of substrate at which the reaction rate is half of its maximum value (V_{max}). The K_m value indicates how well an enzyme can perform its activity at different substrate concentrations; a lower Km value indicates higher affinity between substrate and enzyme and, hence, better performance.

Limitations of the Michaelis-Menten equation

1. The Michaelis Menten equation fails to account for the influence of enzyme inhibitors or activators on the reaction rate.
2. This equation does not consider the allosteric regulation, which is crucial in determining enzyme kinetics.
3. The given equation assumes that the reaction is taking place under constant temperature and pH conditions, neglecting any potential influence of variations in these parameters on the reaction rate.

Lineweaver-Burk Plot

The determination of Vmax and Km from a hyperbolic plot becomes unattainable since Vmax is only reached when the substrate concentration is infinitely high. Lineweaver and Burk proposed transforming the Michaelis-Menten equation into a linear one to overcome this challenge. This transformation led to the creation of the Lineweaver-Burk plot, also called the double reciprocal plot, which graphically represents the Lineweaver-Burk equation in enzyme kinetics. Hans Lineweaver and Dean Burk introduced this plot in 1934 as a valuable tool for studying enzyme kinetics.

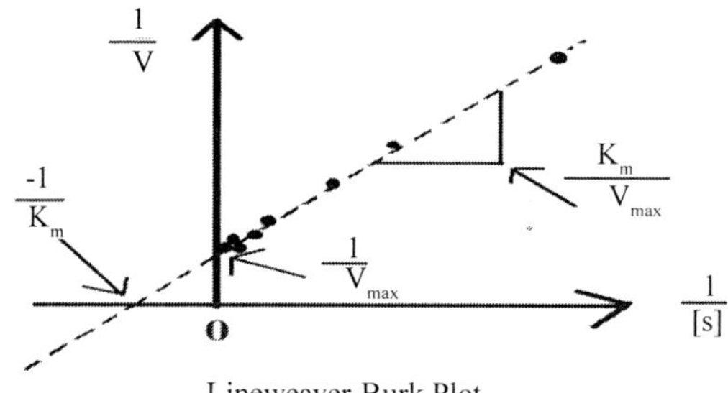

Lineweaver-Burk Plot

The graphical representation presented below is a modified form of the Michaelis-Menten equation.

$$\frac{1}{V} = \frac{K_m + [S]}{V_{max}[S]} = \frac{K_m}{V_{max}}\frac{1}{[S]} + \frac{1}{V_{max}}$$

Where:

V = Reaction velocity

K_m = Michaelis-Menten constant

V_{max} = Maximum velocity

S = Substrate concentration

The Lineweaver–Burk plot is a valuable tool in identifying different types of enzyme inhibition, including competitive, pure, non-competitive, and uncompetitive inhibitors. By analyzing the plot, one can discern the distinct modes of inhibition and compare them to the uninhibited reaction.

Enzyme inhibition

Enzymes possess one or more binding sites like substrate binding sites and/or allosteric sites where substrates or modulators (inhibitors or activators) bind, thereby regulating the rate of reactions. Enzyme inhibitors are compounds that attach to enzymes and can impede the activity of these biological catalysts, leading to a decrease in the overall reaction rate catalyzed by the enzymes. Like enzyme inhibitors sometimes bind to the substrate binding sites, preventing substrates from binding to the active sites and consequently reducing the rates of enzyme-catalytic reactions. The enzyme inhibition can either be temporary or permanent depending upon the nature of the interaction between enzyme and inhibitor. Therefore, based on this, the two main categories of enzyme inhibitors are found: reversible and irreversible inhibitors. Reversible inhibitors bind temporarily (non-covalently) to enzymes; once they are released, the rate of enzyme-catalyzed reactions returns to its normal speed. On the other hand, irreversible inhibitors bind covalently to the enzyme, permanently altering the shape of enzymes and preventing them from forming ES complexes. Thus, irreversible inhibitors are covalent modifiers of an enzyme that ensure the irreversibility of the chemical reaction between the enzyme and the inhibitor, for example, cyanide and Diisopropyl fluorophosphate (DIPF).

Temporary (reversible) enzyme inhibition reactions can be classified into three types.

Competitive inhibition

Competitive inhibition refers to the inhibition of enzymatic activity caused by the competitive binding of inhibitors to the substrate binding site of an enzyme. These inhibitors, referred to as competitive inhibitors, compete with the substrate for binding to the substrate binding site of an enzyme. They possess a similar structure to that of the substrate, enabling them to bind to the enzyme. Instead of forming an enzyme-substrate (ES) complex, the inhibitor forms an enzyme-inhibitor (EI) complex. The substrate (S) and inhibitor (I) concentration influence the inhibition degree. Increasing the concentration of inhibitors intensifies the inhibition and vice versa. While increasing the concentration of the substrate reduces inhibition intensity as high substrate concentrations can successfully displace the inhibitor from the active site.

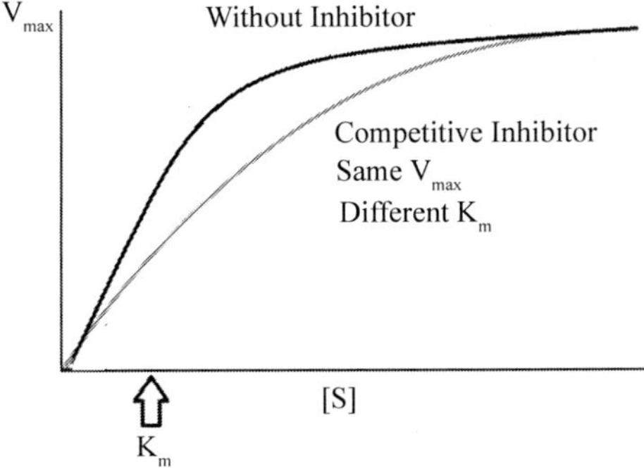

By competing with the substrate at the active site, competitive inhibitors cause an increase in K_m value. However, V_{max} remains unaffected as the reaction can still proceed to completion when there is ample substrate concentration. As a result, the graph depicting enzyme activity against substrate concentration would shift to the right due to the elevated Km. At the same time, the Lineweaver-Burke plot would display a steeper inclination when compared to the scenario without an inhibitor.

Examples of competitive inhibition are as:

1. Succinate dehydrogenase convert succinate to fumarate.
   Succinate-succinate dehydrogenase C Fumarate + NADH + H^+
   Malate is competitive inhibitor of succinate due to structural analogy.
   Malate + NAD+-Succinate dehydrogenase → Oxaloacetate
2. Sulphonamide is competitive inhibitor of PABA during tetrahydrofolate
3. Treatment of methanol poisoning
   Methanol - alcohol dehydrogenase → Formaldehyde (toxic)
   Ethanol - alcohol dehydrohenase → Acetaldehyde

Competitive Inhibition

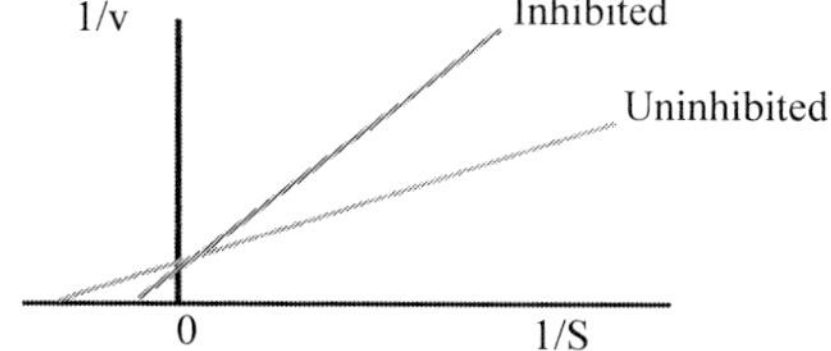

The presence of competitive inhibitors does not alter the apparent value of V. Consequently; competitive inhibitors exhibit the same intercept on the y-axis as uninhibited enzymes. Competitive inhibition either elevates the apparent value of K_m or diminishes substrate affinity. This can be visually observed in the graph as the inhibited enzyme displays a greater intercept on the x-axis.

Non-competitive inhibition

In this inhibition scenario, there is no binding competition between the substrate and inhibitor to the enzyme. Because the inhibitor's binding to a separate site on the enzyme differs from the substrate binding site. As a result, the inhibitor does not affect the enzyme's binding to the substrate because of this distinct binding site. Consequently, increasing the substrate concentration cannot overcome this type of inhibition, thus indicating that changes do not influence the inhibition in substrate concentration. A decrease in Vmax characterizes the enzyme kinetics of this reaction while the K_m value remains constant. Both the substrate and inhibitor bind to the enzyme with equal affinity. However, upon inhibitor binding, a conformational change occurs in the enzyme, preventing the substrate from binding to the enzyme. This ultimately leads to a decrease in Vmax.

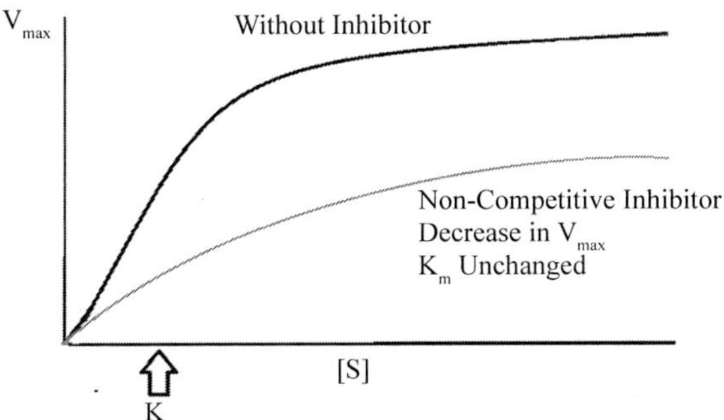

Examples of non-competitive inhibition are as follows:

1. The presence of heavy metals such as Hg^{2+}, Ag^{+}, and Pb^{2+} can lead to heavy metal poisoning, which can distort the enzyme containing the -SH group at the allosteric site. In the cases of metal poisoning, a chelator is employed to mitigate the effects.
2. The proteinase enzyme in bacteria is effectively inhibited by doxycycline, which acts as a non-competitive inhibitor.

Non-competitive inhibition decreases the apparent value of V. This phenomenon is evident on the Lineweaver-Burk plot, where an intercept is increased on the ordinate axis. However, the abscissa intercept (-1/km), which represents substrate affinity, remains unaffected by non-competitive inhibition. Cleland, recognizing the scarcity of non-competitive inhibition in practical applications, identified its occurrence primarily concerning protons and certain metal ions.

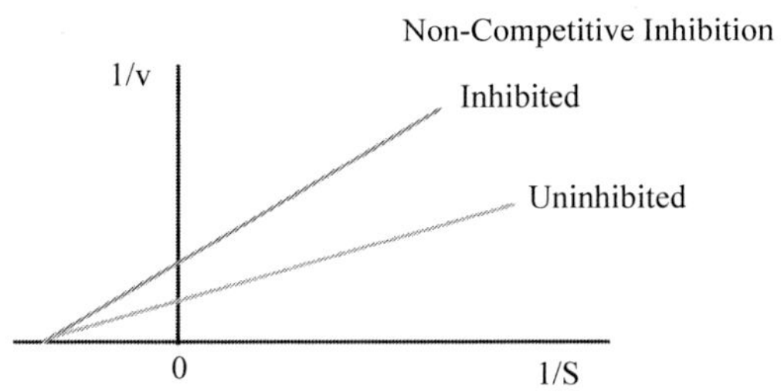

Uncompetitive inhibition

This type of inhibition is observed in multi-substrate reactions and is considered to be relatively rare. The inhibition process shares similarities with non-competitive inhibition, but the inhibitor specifically binds to the ES complex. Initially, the substrate binds to the enzyme, resulting in the formation of the ES complex. Once the substrate is bound to the enzyme's active site, the allosteric site undergoes a conformational change to create a binding site for the inhibitor. Subsequently, the inhibitor binds to the allosteric site and distorts the enzyme-substrate complex, thereby inhibiting catalysis. Based on enzyme kinetics, Vmax and Km decrease in this type of inhibition.

Uncompetitive inhibitors such as tetramethylene sulfoxide and 3-butylthiolene oxide have been found to inhibit liver alcohol dehydrogenase effectively.

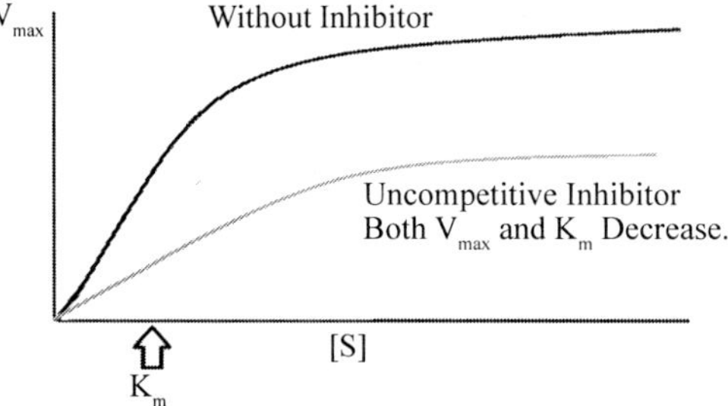

With uncompetitive inhibition, both the apparent values of V and K_m decrease. This can be visually observed on the Lineweaver-Burk plot, where the intercept on the ordinate increases while the slope remains unaltered. Uncompetitive inhibition leads to an increase in substrate affinity or a decrease in the apparent value of K_m. On the plot, uncompetitive inhibition is characterized by parallel lines for different inhibitor concentrations.

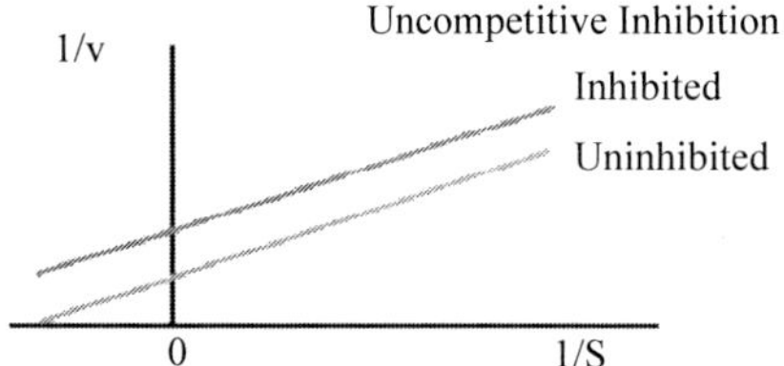

Allosteric enzymes

Allosteric enzymes exhibit an additional binding site for effector molecules, separate from the active site. This binding event triggers conformational changes, thereby modifying their catalytic properties. In addition to the active site or substrate-binding site, allosteric enzymes feature an extra site. This site, known as the C-subunit, is responsible for substrate binding, while the R-subunit or regulatory subunit is responsible for effector binding. Notably, the velocity vs substrate concentration graph of allosteric enzymes deviates from the usual hyperbolic curve and instead exhibits an S-curve pattern. The effector molecule can act as either an inhibitor or an activator. All biological systems are effectively regulated, with a range of mechanisms in our body that oversee and respond to internal and external environmental changes. Whether it involves gene expression, cell division, hormone secretion, metabolism, or enzyme activity, every process is meticulously regulated to ensure proper development and survival. Allostery, the process of enzyme regulation, involves binding at one site and influencing subsequent binding at other sites.

Two distinct forms of allosteric regulation are based on the substrate and effector molecules involved. The first type is known as homotropic regulation, where the substrate molecule also acts as an effector. This type is primarily associated with enzyme activation and is commonly referred to as cooperativity. A classic example of this is the binding of oxygen to hemoglobin. On the other hand, heterotropic regulation occurs when the substrate and effector molecules are different. In this case, the effector can either activate or inhibit the enzyme. An example of this is the binding of CO_2 to hemoglobin.

There are two types of allosteric regulation based on the action performed by the regulator: inhibition and activation. Allosteric inhibition occurs when an inhibitor binds to the enzyme, causing conformational changes in all the active sites of the protein complex. As a result, the enzyme's activity decreases. In simpler terms, an allosteric inhibitor is a molecule that specifically binds to an allosteric site on the enzyme. On the other hand, allosteric activation happens when an activator binds to the enzyme, enhancing the function of the active sites and leading to an increased binding of substrate molecules.

Two primary models can explain the cooperativity in allosteric enzymes.

1. Concerted model

Monod, Wyman, and Changeux initially introduced the concerted model, also referred to as the MWC model, in 1965. As per this model, enzymes comprising multiple subunits can exist in two conformational states: the relaxed (R) state and the tense (T) state. These states differ in the energy of their interactions between subunits while maintaining symmetry, implying that all subunits are either in the T-state or the R-state. In the absence of ligands or substrates, the enzyme predominantly adopts the T-state. However, upon ligand binding, the protein simultaneously transitions from the T-state to the R-state in all the subunits without forming a hybrid RT state.

2. Sequential model

In 1966, Koshland, Nemethy, and Filmer proposed the sequential model, also called the KNF model. According to this model, the binding of a ligand to the enzyme triggers a conformational change exclusively in the subunit to which the ligand binds. This alteration subsequently influences the neighboring subunits, leading to cooperative behavior. Unlike the MWC model, the KNF model does not maintain symmetry during the binding process as it encompasses a combination of both R and T states. Subunits undergo conformational change sequentially, resulting in the formation of an RT state.

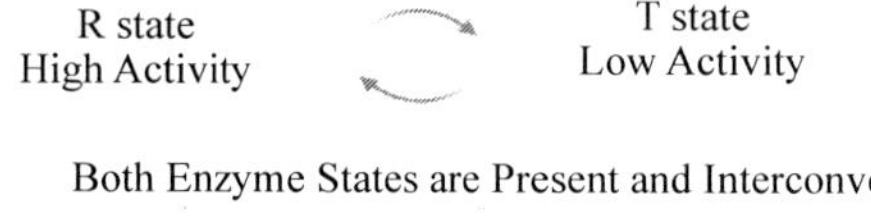

Both Enzyme States are Present and Interconvert in Solution

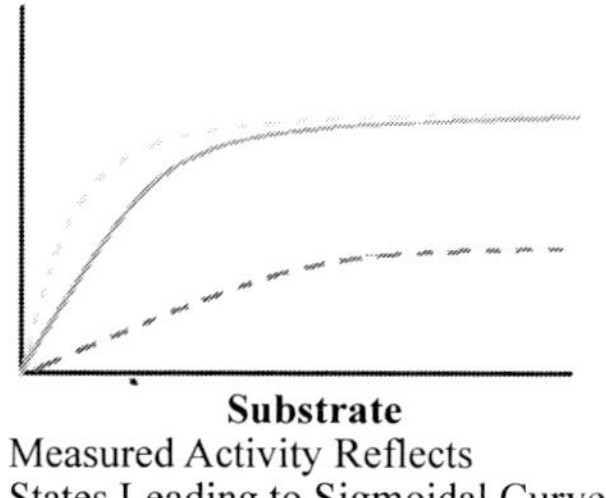

Measured Activity Reflects States Leading to Sigmoidal Curve

Activators

Inhibitors

Activators Push Towards R State
Inhibitors Push Towards T State

Examples of allosteric inhibition

1. Aspartate transcarbamoylase (ATCase)

ATCase, an enzyme crucial for the synthesis of pyrimidine nucleotides, which are the fundamental units of DNA and RNA, plays a significant role. Among the end products of this pathway, Cytidine triphosphate (CTP) acts as an allosteric inhibitor. The binding of CTP to the allosteric site of ATCase induces a structural alteration in the enzyme, reducing its activity. This feedback inhibition mechanism effectively prevents the excessive accumulation of pyrimidine nucleotides.

2. Phosphofructokinase (PFK-1)

PFK-1, a pivotal enzyme in the glycolysis pathway, plays a vital role in the breakdown of glucose to produce energy. Notably, PFK-1's activity is regulated by the presence of ATP, which acts as an allosteric inhibitor. When ATP molecules bind to the allosteric site of PFK-1, its activity is inhibited, ensuring excessive glucose breakdown is avoided when the cell has abundant energy. This mechanism helps maintain a balanced and controlled glycolytic process.

Functions of enzymes

The functions of enzymes within our bodies are diverse and essential, and some of them are mentioned below:

- One such role is their involvement in signal transduction, where the enzyme protein kinase catalyzes the phosphorylation of proteins.

- Additionally, enzymes aid in the breakdown of large molecules into smaller substances, facilitating their absorption by the body.
- Furthermore, enzymes are crucial in energy generation, with ATP synthase being the specific enzyme responsible for energy synthesis.
- Moreover, enzymes are responsible for facilitating the movement of ions across the plasma membrane.
- In addition to these functions, enzymes perform various biochemical reactions, including oxidation, reduction, and hydrolysis, which contribute to eliminating non-nutritive substances from the body.
- Lastly, enzymes play a vital role in reorganizing the internal structure of cells, thereby regulating cellular activities.

Summary

- Enzymes play a crucial role in facilitating chemical reactions within the body. Their primary function is to accelerate the rate of these reactions, thereby contributing to the sustenance of life.
- The body's enzymes are involved in a wide array of vital tasks, such as muscle development, detoxification, and the breakdown of food particles during the process of digestion.
- Amylase, protease, and peroxidase are among the plant-derived enzymes that can be found in various plant species.
- Regarding their chemical nature, enzymes are generally recognized as globular proteins. However, it is important to acknowledge that ribozymes, a specific type of RNA molecule, can also possess enzymatic properties. These ribozymes primarily serve the purpose of cleaving other RNA molecules.
- Several factors, including enzyme concentration, substrate concentration, temperature, pH, and salt concentration, can impact the activity of enzymes.
- A small crevice is commonly observed on the enzyme's surface, serving as an active or catalytic site. This site facilitates the binding of one or two specific substrates.
- The process of allosteric inhibition involves the binding of a molecule to an allosteric site on an enzyme, causing a conformational change in the active site. This alteration obstructs the entry of the substrate, effectively ceasing the enzyme's activity.
- In allosteric activation, a regulatory molecule binds to the secondary (allosteric) site, inducing a conformational change in the active site.

This conformational change enables the substrate to bind, ultimately resulting in the synthesis of the product.

Long Type Questions

1. Describe enzymes along with their classification.
2. Write an elaborate note on enzyme inhibition and its types.
3. What are the various factors that affect enzyme activity?
4. Describe the mechanism of enzyme action.
5. Describe briefly the Michaelis-Menten equation along with its limitations.
6. Write a short note on the Lineweaver-Burk Plot and its applications.
7. What is allosteric inhibition? Describe its different inhibition models.
8. Write a short note on the functions of the enzymes.

Fill in the Blanks

1. Enzymes are __________ that play a crucial role in accelerating chemical reactions without undergoing any changes themselves.
2. The proteinous enzymes are classified into _________ and __________ enzymes.
3. Enzymes require the presence of non-proteinous substances called ___________ to carry out their functions effectively.
4. Some examples of prosthetic groups are _________, _________, __________ and ____________
5. The International Union of Biochemists (IUB) has classified enzymes into ________ distinct functional classes.
6. Scientists Leonor Michaelis and Maud Leonora Menten formulated the Michaelis-Menten equation in the 20th century to explain the correlation between ________________ and the rates of enzyme-catalyzed reactions.
7. Competitive inhibitors cause an __________ in K_m value with V_{max} remains__________
8. In non-competitive inhibition, there is a _________ in _________ where as _________value remains ___________
9. In uncompetitive inhibition, both V_{max} and K_m ______________
10. Allosteric inhibition occurs when an inhibitor binds to the enzyme, causing conformational changes in all the ___________of the protein complex.

8

Vitamins

Vitamins are natural and indispensable nutrients that are essential for our bodies in small quantities. Vitamins are indispensable for a multitude of biological functions, including growth, development, wound healing, maintaining healthy bones and tissues, supporting the immune system, and more. These vital organic compounds exhibit diverse biochemical functions and are essential for metabolic processes. The discovery of vitamins dates back to 1912 when Casimir Funk, a Polish American biochemist, initiated research on their sources, functions, and deficiency disorders. Because of his groundbreaking work, Funk is widely recognized as the pioneer of vitamins.

Types of vitamins

The classification of vitamins is determined by their solubility, resulting in two distinct groups: fat-soluble vitamins and water-soluble vitamins.

Fat-soluble vitamins

Fat-soluble vitamins, as their name implies, need fat for absorption and are stored in the fat cells. These vitamins include Vitamin A, D, E, and K.

Vitamins	Chemical name	Food sources	Functions	Deficiency
Vitamin A	Retinol	Carrots, tomatoes, green vegetables, liver oil, milk, fish, butter, spinach, pumpkin, eggs.	Vitamin A plays a crucial role in maintaining healthy vision, supporting a strong immune system, promoting proper growth and development of embryos, and ensuring the well-being of the skin and mucous membranes.	Night blindness (Nyctalopia) Xeropthalmia, skin keratinization
Vitamin D	Calciferol	Fish, liver, egg beef, chicken breast, cod, sunlight exposure, etc.	Vitamin D is crucial in maintaining the balance of calcium and phosphorus levels within the body.	Rickets and Osteomalacia
Vitamin E	Tocopherol or alpha-tocopherol	Fortified cereals, leafy green vegetables, seeds, potatoes, pumpkin, guava, mango, milk, and a variety of nuts and seeds.	Vitamin E has proven to be highly efficient in preventing and reversing a wide range of disease complications. This is primarily attributed to its antioxidant properties, ability to combat inflammation, capacity to inhibit platelet aggregation, and its immune-enhancing activity.	Heart problems, hemolysis, and sterility
Vitamin K	Antihemorrhagic factor	Tomatoes, broccoli, grapes, cashews, mangoes, chestnuts, nuts, beef, and lamb,	Vitamin K is crucial in supporting the body's blood clotting process and promoting healthy bone development.	Excessive bleeding, inadequate bone growth, osteoporosis, and heightened risk of cardiovascular ailments.

Water-soluble vitamins

Water-soluble vitamins are not stored in the body but are excreted through urine when consumed in excess. As a result, it is essential to replenish these vitamins to maintain optimal health continuously. Water-soluble vitamins include B and C.

Vitamins	Chemical name	Food sources	Functions	Deficiency
Vitamin B1	Thiamine	Whole grains, legumes such as beans, green peas, lentils, oatmeal, sunflower seeds, soymilk, yogurt, and fortified foods.	Thiamine assists in the conversion of carbohydrates into energy within the body's cells. Furthermore, it is involved in the regulation of muscle contraction and the transmission of nerve signals.	Beriberi
Vitamin B2	Riboflavin	Fruits and vegetables include bananas, dates, mushrooms, grapes, mangoes, peas, pumpkin, and popcorn.	Maintaining the body's energy supply supports growth and development, strengthens the immune system, and promotes skin and hair health.	Fatigue, dermatitis, blurred vision, swollen throat, skin crack, and hair loss.
Vitamin B3	Niacin	Meat, eggs, mushrooms, fish, cereals, guava, and dairy products.	Vitamin B3 plays a vital role in maintaining the optimal functioning of the gastrointestinal system. It enhances digestion, facilitates the proper absorption of food juices into the body, and aids in the elimination of waste products from the intestines. Additionally, it effectively prevents constipation and alleviates flatulence.	Pellagra
Vitamin B5	Pantothenic acid	Plant and animal sources include meat, vegetables, cereal grains, legumes, eggs, and milk.	Vitamin B5 is essential for the production of energy, red blood cells, and sex hormones.	Fatigue, depression, insomnia, burning feet, vomiting, irritability, and stomach pains,
Vitamin B6	Pyridoxine	Poultry (chicken or turkey), soya beans, wheat germ, peanuts, snacks, and oats.	It helps in the formation of hemoglobin and red blood cells, maintains normal nerve function, and reduces fatigue and tiredness.	Microcytic anemia, dermatitis, and peripheral neuropathy.

Vitamin B7	Biotin or Vitamin H	Organ meats, eggs, fish, seeds, nuts, and specific vegetables like sweet potatoes	Biotin is an essential component involved in diverse metabolic processes, primarily focused on efficiently utilizing fats, carbohydrates, and amino acids. Furthermore, it regulates immune and inflammatory functions within the body. Biotin's significance extends to promoting healthy eye, hair, skin, and brain function. Additionally, it plays a vital role in maintaining the optimal functioning of various organs, including the eyes, kidneys, and even during pregnancy.	Depression, lethargy, dermatitis, tingling sensation in the extremities, numbness, hallucinations, nausea, hair loss, and ataxia.
Vitamin B9	Folic acid	Liver, dairy products, egg yolk, seafood, Dark green leafy vegetables, beans, peanuts, sunflower seeds, fresh fruits, fruit juices, whole grains, wheat germ, cereals, breads, and pasta.	Production of genetic material, formation of red blood cells, prevention of anemia and neural tube malformation.	Megaloblastic anemia
Vitamin B12	Cobalamin or cyanocobalamin	Fish, meat, poultry, eggs, milk, and other dairy products	Supports the vitality of blood and nerve cells, contributes to DNA synthesis, and prevents megaloblastic anemia.	Pernicious anemia and neurological problems
Vitamin C	Ascorbic acid	Citrus fruits such as oranges, kiwi, lemon, grapefruit, bell peppers, strawberries, tomatoes, cruciferous vegetables like broccoli, brussels sprouts, cabbage, and cauliflower, as well as white potatoes	Collagen formation, antioxidant properties, growth, and repair strengthen the immune system and wound healing.	Scurvy

Summary

- Vitamins play a crucial role in maintaining optimal metabolic function as they are vital organic compounds.
- Casimir Funk is credited with coining the term "Vitamin" in 1912.
- Vitamins are essential for various biological functions, including growth and development, wound healing, bone and tissue health, and the proper functioning of the immune system.
- Vitamins are commonly divided into two groups: fat-soluble and water-soluble.
- Fat-soluble vitamins, including A, D, E, and K.
- Water-soluble vitamins include B and C.
- Deficiency disorders occur when there is a dearth or inadequacy of vitamins.
- There are three distinct forms of vitamin A, which include retinol, retinal, and retinoic acid.

Long Type Questions

1. Write a short note on vitamins and their types.
2. Explain fat-soluble vitamins in detail.
3. Describe water-soluble vitamins in detail.

Fill in the Blanks

1. Vitamins are organic compounds required in ______________ quantities.
2. Vitamins are classified into ______ types.
3. Fat-soluble vitamins include ______, ______, ____, and ______
4. Water-soluble vitamins include _________ and ________
5. The deficiency of vitamin D causes __________ and __________
6. ____________ is also known as an antihemorrhagic factor.

Solutions to Chapters

Chapter 1

1. Passive transport
2. Active transport
3. Passive transport
4. Pinocytosis, phagocytosis

5. Exocytosis
6. Acidic
7. Acidic, basic
8. Weak acid, conjugate base
9. Henderson-Hasselbalch equation

Chapter 2

1. Aldehydes, Ketones
2. $C_n(H_2O)_n$
3. Monosaccharides, Oligosaccharides, Polysaccharides
4. Sugars, Cannot
5. Complex, 2-10
6. Ten, Glycosidic linkage
7. Starch, glycogen
8. Cellulose, Chitin

Chapter 3

1. Wilhem Bloor, 1943.
2. Unsaturated
3. Waxes
4. Derived
5. Eicosanoids

Chapter 4

1. Proteins
2. Acidic, basic, side chain, hydrogen
3. Eight
4. Hydrophilic
5. Arginine, Histidine, Lysine
6. Glutamate, Aspartate
7. Non-essential amino acids
8. Essential amino acid
9. Glycine

Chapter 5

1. Amino acids
2. Peptide bonds
3. Amino (NH_2), Carboxyl (COOH)
4. Two, Ten
5. Fibrous, Scleroprotein
6. Blood plasma, Globular
7. Five, Lipoproteins

Chapter 6

1. Nucleotides.
2. Uracil.
3. 5' –> 3'
4. Hydrogen
5. Thymine
6. Friedrich Miescher

Chapter 7

1. Biocatalysts
2. Simple, conjugated
3. Cofactors
4. Heme, copper, biotin, ubiquinone
5. Six
6. Substrate concentration
7. Increase, unaffected
8. Decrease, V_{max}, K_m, constant.
9. Decreases
10. Active sites

Chapter 8

1. Small
2. Two
3. A, D, E, K
4. B and C
5. Rickets and Osteomalacia
6. Vitamin K

Index